The
International Mars
Research Station

*An exciting new plan
to create a permanent
human presence on Mars*

Shaun Moss

First published in March 2015.

ISBN-13: 978-1508683940
ISBN-10: 1508683948

Library of Congress Control Number: 2015903907
CreateSpace Independent Publishing Platform, North Charleston, SC

marsbase.org

This book is dedicated to Olive Kalani Moss, my beautiful niece.

May you dance under Martian skies!

Contents

Acknowledgements

There are a few people I should thank for various reasons. My deepest gratitude is extended to the following...

Mum, Dad and Alex for always being there for me. I am lucky to have such a great family.

The guys and gals from the Mars Society, Mars Society Australia, Mars Settlement Research Organisation and Space Development Steering Committee, who have all been brilliant. The book has benefited greatly from the insightful feedback, enthusiasm, support, sharp intellects and visionary imaginations of fellow areophilic scientists, engineers, artists and space revolutionaries.

Jonathan Clarke and David Willson, President and Vice President respectively of Mars Society Australia, for helping me with several aspects of the architecture and encouraging me when I needed it. Thanks especially to Jon who suggested the excellent location of Lyot.

William Mook and Adam Crowl for helping with the rocket science, Dava Newman for answering my questions about the BioSuit, James Fesmire from the Cryogenics Test Laboratory at NASA KSC, who helped with my questions about multilayer insulation, Drago Ilas for his useful delta-V map of the Solar System, and Tom Gangale for his creative insights into efficient Earth-Mars communications.

The organisers of TEDx Noosa 2014 for allowing me the opportunity to present this idea.

My awesome proofreaders and friends who provided such valuable feedback: Jonathan Clarke, Claire Lentz, Rob Parker, Conrad Steenkamp, Babak Shakuri, Pekka Lehtikoski, Isaac Arthur, Andrew McKeikan and Tom Gangale.

And last, but definitely not least, the luminaries of the Mars movement from whom I have drawn so much inspiration: Dr Robert Zubrin, Dr Christopher McKay, Kim Stanley Robinson, Elon Musk, and other great teachers and leaders. Without your shining light and vision, I might not have seen nor believed.

Preface

The writing and design project that became this book began in approximately April of 2013 while I was living in Thailand. The advent of the SpaceX Falcon 9 and Dragon capsule, and the Bigelow Aerospace B330 (then the "BA 330"), led me to begin formulating ideas for a Mars mission architecture that I believed had certain important advantages over existing proposals. I started writing up the ideas to share with friends online, then created the Facebook group "Mars Settlement Research Organisation" as a way to discuss the plan, and Mars settlement generally, with the online community. Therefore, even though IMRS (International Mars Research Station) is only one of hundreds of topics discussed by MSRO (now with over 6000 members), in my history they are strongly linked and have evolved in parallel.

The seeds of books are usually sown long before any actual writing starts, and such is the case with this one. I've been interested in settlement of Mars and a participant in the Mars community ever since reading *Red Mars* by Kim Stanley Robinson and subsequently joining the Mars Society in 1999, and since that time have been learning about, and developing ideas for, human exploration and settlement of the red planet.

Working on this project led to my TEDx talk about IMRS and its architecture, in April 2014. That talk, and other presentations and interviews by me, can be found at my YouTube channel (https://www.youtube.com/user/mossy2100/videos).

The book has some shortcomings, for which I apologise. For starters, this is an exercise in aerospace engineering being attempted by a computer scientist, which is something like a mechanic trying to build a house. Many aspects of the plan and designs are less developed than they would be by someone with more experience or resources. Thus, rather than a fully completed plan, this work is offered simply as a group of related ideas that will hopefully interest or inspire others with greater skill and experience. The intention is to trigger further discussion.

The layperson may find that the book is overly reliant on maths, acronyms and chemical symbols. I have tried to ease the pain of these inclusions as much as possible; however, it's difficult to write a book like this while excluding them altogether. Calculations can safely be skipped over without loss of meaning; they're included mainly for the benefit of friends of mine in the Mars community who will wonder or ask how I worked out this or that result. As for acronyms and initialisms, they're an inherent feature of space community dialogue, and hard to avoid without being overly verbose.

I couldn't include everything in the book that I would have liked, otherwise you wouldn't be reading this now; it would still be living within the virtual confines of Scrivener. This project had no clearly definable end — other than, perhaps, a permanent human presence on Mars — and I could have happily expanded on any one of the subtopics or designs presented here with progressively greater detail *ad infinitum*. Further complicating matters, science and technology are advancing so rapidly now that new ideas for improving the plan suggest themselves almost weekly. When explaining this problem to friends, 1.5 years into the project, they all suggested saving any further ideas for a future edition or sequel, which I accepted as good advice.

Despite these imperfections, I hope you will enjoy the book. I believe the principles underlying the development of the mission architecture, the plan to settle Mars as a truly international effort, and at least some of the vehicle design concepts, offer a few unique contributions to the evolution of our quest to put human bootprints on Mars.

For the Mars-layperson, I hope the book will provide an effective introduction to many aspects of the challenge of sending people to the red planet and bringing them safely home. I think almost anyone can learn something about Mars from this book.

I've used British English and the metric system throughout; sorry, America. During the coming year I will try to have the book translated into the languages of the top space-faring nations.

Like any good software developer, I would like to solicit user feedback that may help to inform future efforts. Therefore, if you have any comments or suggestions, please submit them shaun@astromultimedia.com or message me on Skype, Facebook or Twitter (username is mossy2100 everywhere). You may also like to join the Mars Settlement Research Organisation or the Mars Society.

With much love and gratitude,

Shaun Moss

March 2015

About This Book

Nomenclature

Some words about words are necessary.

1. The word "Earthian" is used as the proper noun referring to things of Earth. Although perhaps not as common in mainstream literature, particularly science fiction, as "Terran", it is often preferred in academic writing. (The term was also oft-used by one of the 20th century's greatest geniuses, R. Buckminster Fuller.) It is also a more suitable word than "terrestrial". Although "extraterrestrial" means "not from Earth", "terrestrial" is more usually an antonym of "aquatic", i.e. meaning "of the land" rather than "from Earth".

2. The equivalent proper noun for Mars is "Martian".

3. The words "Earthian" and "Martian" are always capitalised (e.g. "she is a Martian", "the Martian sky"), being proper nouns analogous to "Australian", "Italian", etc.

4. The words "Earth", "Moon" and "Sun" are also proper nouns and therefore always capitalised.

5. In accordance with modern usage, the word "Earth" is not preceded by the definite article "the", unlike "the Moon" and "the Sun".

6. When talking about Mars, the prefix "areo", meaning "Mars", replaces "geo", meaning "Earth", in words that normally begin with that prefix. Thus, "geothermal" becomes "areothermal", "geology" becomes "areology", "geostationary" becomes "areostationary", and so forth.

7. The unusual words "periareon" and "apoareon" refer to the low and high points, respectively, in an orbit around Mars. They are Martian equivalents of the words "perigee" and "apogee", which refer to orbits around Earth, and "perihelion" and "aphelion", which refer to orbits around the Sun.

8. The word "methalox" refers to bipropellant comprised of liquid oxygen and liquid methane. There's no standard term for this, and variants include LO2/LNG, LOX/LCH4, LOx/Methane, and every possible combination and permutation. Using the word makes the text more readable and reduces acronym density.

9. As discussed in <u>Timekeeping</u>, the word "sol" appears frequently where you might normally expect to read "day", because the crew will be operating on Mars time. However, the words "day" and "night" are still used for the light and dark parts of the sol, respectively. No special word is used for a Martian week, which is a period of 7 sols.

10. For conciseness, the initialism "EVA", which stands for "Extra Vehicular Activity", refers to any activity outside a vehicle or habitat during which the astronaut is wearing a spacesuit. Although the initialism "EHA" for "Extra Habitat Activity" is sometimes used, "EVA" is much more common and often refers to working outside a habitat, even if it's technically not a vehicle.

Acronyms and initialisms

There's a large number of acronyms and initialisms in this book, drawn primarily from usage in space, science and engineering, plus a few from IT and military, and some have been newly invented for the purpose of this document. Unfortunately this is the nature of the topic, and acronyms are a preferable shorthand to the expanded version in most cases. Each is expanded on first use, and can also be looked up in the <u>Abbreviations</u> section towards the end of the book. Abbreviations for chemical formulae are also provided in the same section.

Updates

In response to feedback and corrections received after the book goes live, or if I feel inspired to add new material, I may make updates and re-upload the book. In addition, if the right people like the idea, there could be progress with regard to actual implementation of the IMRS plan.

If you would like to be kept informed of updates like these, please sign up to the newsletter at <u>marsbase.org</u>.

Introduction

Welcome to the International Mars Research Station, a plan to bring the world's leading space-faring nations together to cooperate in the establishment of a permanent human presence on Mars for the benefit of all humanity.

This book presents a collection of ideas related to a program of near-term, affordable and achievable missions to send multiple international crews to Mars and return them safely to Earth, while accumulating hardware on Mars and developing a permanent base. The intention is an initial step towards settlement. The concept for the base is a shared facility developed by a partnership comprised of the world's top space agencies, for use by, and for the benefit of, all nations and peoples.

Many plans for sending humans to Mars have been proposed over the years, with each iteration incorporating new innovations and technologies, filling gaps and solving problems, and bringing us step-by-step closer to achieving the goal. It is now widely believed within the Mars community that there remain few, if any, genuine technological barriers to sending humans to Mars, and that almost all elements required for a successful mission are now available either from commercial suppliers, or are being, or could be, developed by government space agencies or research institutions in the relatively near future. Although some crucial mission components may still need to be developed, the primary challenges are now arguably financial and political rather than technological.

Perhaps the most important and disruptive architecture presented in recent years was Mars Direct (Zubrin, Baker & Gwynne 1991), principally because it introduced the idea of ISPP (In Situ Propellant Production) as a method for drastically reducing the mass to be transported to Mars, and therefore the investment required for a successful HMM (Humans Mars Mission). This represented a quantum leap forward compared to existing HMM concepts.

NASA (National Aeronautics and Space Administration) suggested some improvements to Mars Direct, which led Dr Zubrin to develop an evolution of the architecture known as "Mars Semi-Direct". NASA then developed a new Mars mission architecture based on Mars Semi-Direct, known as the NASA Design Reference Mission. This architecture has evolved to version 5.0 and is now called the Design Reference Architecture (ed. Drake 2009). The DRA has been in development for at least 12 years, involving many experts and departments within NASA, and undoubtedly represents one of the best and most evolved HMM architectures currently available.

Nonetheless, improvements to the DRA continue to be proposed, suggesting further opportunities for its evolution. A few examples:

- The use of atmospheric water to replenish supplies during the surface mission (Grover et al. 1998).

- Biconic spacecraft to improve EDL (Entry, Descent and Landing) performance (Willson & Clarke 2005).

- Horizontal rather than vertical cylinders for habitats in order to more efficiently optimise geometry (Willson & Clarke 2005).

- Use MLLV (Medium Lift Launch Vehicles) or EELV (Evolved Expendable Launch Vehicles) to circumvent the significant investment required to develop a suitable SHLLV (Super Heavy Lift Launch Vehicle) (Bonin 2006).

- Adopting a "base-first" approach rather than a series of missions to disparate locations (Gage 2010).

- Inflatable modules to significantly expand habitat volume (Kozicki & Kozicka 2011).

- Capsules to safely deliver cargo and crew (Wielders et al. 2013).

- The use of SEP (Solar Electric Propulsion) for Earth-Mars travel (Raftery et al. 2013).

At the heart of the IMRS plan is a modern, practical, safe (or, at least, safer) and more affordable architecture for human Mars missions, which makes use of the latest advancements in space technology. This architecture, called "Blue Dragon", is largely an evolution of the DRA, incorporating innovations designed to improve cost-effectiveness, safety, reliability, the likelihood of success, and the overall benefit to humanity. It takes advantage of a number of COTS (Commercial Off-The-Shelf) hardware components that have recently become or will soon become available. The result is an affordable and achievable architecture with improved outcomes.

One of the main differences from the DRA is the choice to focus on a single location rather than several, thereby enabling reuse of hardware elements across multiple missions. Another important feature is the use of reusable rockets, capsules and habitats. ISRU (In Situ Resource Utilisation) technology is leveraged for manufacturing propellant, breathable air and water, thus reducing launch mass and mission cost.

These strategies for controlling program cost will facilitate sending humans to Mars sooner, and will enable more missions to be conducted and more countries to be involved, thus benefiting more people.

The mission achieves improved safety by using capsules to ferry crews between the surfaces and orbits of both Earth and Mars; pre-deploying virtually all the base hardware, including the MAV (Mars Ascent Vehicle), surface habitat, surface vehicles, power systems and ISRU equipment; and providing redundant backup systems to the base (e.g. additional habitat, vehicles, solar panels, etc.)

Two concepts for precursor missions are "Green Dragon" and "Gold Dragon", designed to test mission-critical elements of Blue Dragon. These missions are described briefly towards the end of this document.

Aside from architectural changes, there are two other important differences from the DRA related to the intention of the mission:

1. The IMRS is intended as a first step in a broader process of settlement rather than a standalone Apollo-style "flags and footprints" program of purely science and/or exploration. The focus is more on laying the foundations for a permanent human presence: making use of local resources, building up technological infrastructure, studying effects on human physiology and psychology, and generally learning how to live on Mars. This affects the destination as well as surface activities. The fundamental philosophy is to develop a permanent presence as a first priority, then as much science as is desired can proceed much more easily, cheaply, efficiently and conveniently. Research activities during surface stays are therefore primarily oriented around exploration, science and engineering in the context of ongoing habitation of Mars.

2. This plan is not designed to serve the interests of any one nation, but humanity as a whole. For this reason, and also due to the cost and scale of the program, the intention is that it will be completed by an international consortium of space agencies, companies and academic/research institutions. This is considered necessary, not only for budgetary, but also political, ethical and philosophical reasons. If Mars is to be a future home for humanity, it should be for all humanity, and not any single Earthian nation. To become multiplanetary, nationalistic ideas must be left behind. Humanity should explore and settle Mars as a united people.

It is hoped you will enjoy learning more about this exciting project, which has the potential to bring the world together in a peaceful and tremendously historic adventure: opening up a new world for human civilisation.

HERE PEOPLE FROM EARTH
FIRST SET FOOT ON MARS

JULY 2033 CE

WE CAME IN PEACE FOR ALL HUMANITY

Plaque for the first Mars landing based on the plaque for the first Moon landing

1. The IMRS

The DRA proposes a series of three missions to distinct locations on Mars over a period of a decade; in other words, a similar program to Apollo. However, if settlement of Mars is the primary goal, rather than purely exploration, scientific research or political oneupmanship, then an Apollo-style program makes less sense. If we're going to Mars to stay, a superior strategy is to build up infrastructure in one location and establish PHP-M (Permanent Human Presence on Mars) as quickly as possible. This is referred to as a "base first" strategy.

Once a Mars base is established it will be easier to conduct whatever further science and exploration is desired, as expeditions can be launched from the base rather than from Earth. This will be more efficient, cheaper and much more convenient.

For this reason, the program outlined here describes a series of missions to one specific location on Mars. The intention is to progressively build up infrastructure and technological capability at this location, establishing a base that can support multiple missions and be developed to a point where it can support a human presence more-or-less indefinitely. This is the International Mars Research Station (IMRS).

1.1. Shoulders of Giants

As the name suggests, the IMRS is largely inspired by, and draws on the legacy of, the ISS (International Space Station) and the MARS (Mars Analogue Research Station) programs. It also builds on the experience and knowledge garnered through visionary and pioneering space programs such as Apollo, SkyLab, Mir and others.

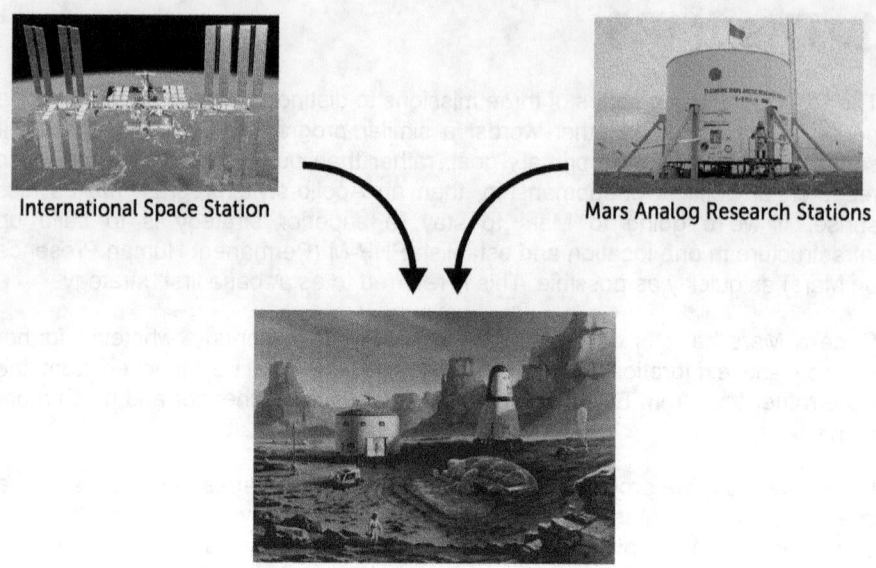

International Space Station

Mars Analog Research Stations

International Mars Research Station

ISS + MARS = IMRS (Credit: NASA, The Mars Society)

1.1.1. The International Space Station

The ISS is a triumph of space engineering and international collaboration, and is one of the greatest success stories in the history of space development. A total of 75 missions have been conducted to the ISS, and it has been permanently inhabited for 14 years. It is also the most expensive artefact ever constructed, with an estimated total cost of US$150 billion. This high cost is one of the reasons why an international partnership has been helpful, as it spreads the cost across multiple countries.

The ISS has also been profoundly important for encouraging and developing peaceful international cooperation in space. This is important partly because it involves cutting-edge technologies that may have military applications; therefore, the collaborative relationships between ISS partners indicates a high level of trust.

The ISS continues to benefit people all over the world, not only the participating nations. It's a beacon of inspiration and achievement at the forefront of human expansion into space. Because the IMRS is designed to be international, it makes sense to continue building on the collaboration framework already developed for the ISS.

1.1.2. Mars Analogue Research Stations

For some years the Mars Society has conducted a program to build and operate several MARSes. To date, two operational MARSes have been constructed by the Mars Society and are in active use:

1. FMARS (Flashline Mars Arctic Research Station), located at Haughton Crater on Devon Island in the Canadian arctic.

2. MDRS (Mars Desert Research Station), located near Hanksville in Utah, USA.

NASA have also built a MARS on the slopes of the Mauna Loa volcano in Hawaii called HI-SEAS (Hawaii Space Exploration Analog and Simulation).

Another, MARS-Oz (Australian Mars Analogue Research Station) is planned for the Lake Frome Plains east of Arkaroola in South Australia, possibly to be constructed within the next 1-2 years.

The MARS program has been enormously successful. To date, 14 simulated missions have been operated at FMARS and more than 150 at MDRS. Mission durations range from weeks to months. Although FMARS was constructed first, fewer missions have been conducted there due to its being located in a much more extreme and less accessible environment. It only operates during the summer, whereas MDRS is operated year-round.

The Mars Society has begun preparation for an ambitious 365-day simulated mission to FMARS, called MA365 (Mars Arctic 365). A refurbishment mission has already been completed, and the crew is currently being selected. The full 365-day simulation will run from 2015-2016.

The Mars Society's and NASA's Mars analogue research programs have enabled them, as well as other space agencies and groups, universities and independent researchers, to gain deep insight into almost every aspect of the surface element of HMMs. This includes EVA; geological and biological research; marssuit and airlock design; habitat layout and operation; recycling and resource management; psychology and human factors; art and education; construction and building materials; food preparation and production; command, control and communications; surface mobility and robotics; navigation; scheduling and rostering; and much more.

Crews who have been involved in simulations at FMARS and MDRS invariably speak highly of the experience. One of the most interesting features is the reuse of a habitat that contains the imprints of previous crews - signatures and inspirational quotes, photographs, furniture, musical instruments, cooking equipment, and the countless tiny tweaks and refinements to improve the interior environment and make it progressively safer and more agreeable as a living space. This is one of the reasons for choosing to establish an outpost at a single location on Mars instead of exploring discrete locations.

Testing robots at FMARS (Credit: The Mars Society)

Heading home after a long day's exploration at MDRS (Credit: The Mars Society)

Looking out over HI-SEAS (Credit: NASA)

1.2. Budget

Cost is a major factor in any space venture, and in light of increasing pressure to allocate public money to more immediate issues, it has become an even more important consideration. NASA's nominal benchmark for implementing the DRA is up to $100 billion for the full program lifecycle, including all development costs and three missions (Larson & Pranke 1999). However, for the first three missions to the IMRS it should be possible to achieve a total cost well below this, for several reasons:

1. The mission architecture is largely assembled from commercially-available (COTS) hardware components, which are considerably cheaper to purchase than it would be to develop custom-designed components.

2. One of the major investments to enable humans-to-Mars missions is an SHLLV capable of delivering at least 20-30 tonne payloads to the surface of Mars. However, with the Space Launch System, Mars Colonial Transport, and Long March 9 all currently in development, a suitable vehicle should be available, and thus the cost of developing one does not need necessarily to be incorporated into the Mars program cost.

3. All missions are to the same location, therefore major components such as the MTV (Mars Transfer Vehicle), SHAB (Surface HABitat), power systems, greenhouses, surface vehicles and so forth can be reused by

multiple crews. The architecture is specifically organised for component reuse across multiple missions.

4. As an international mission, some of the engineering, fabrication and other work can be done by countries where costs of labour, rent and other overheads are lower.

The goal should not be to do the cheapest possible HMM, as overly aggressive cost-cutting measures would compromise safety and results. The goal should be to do the smartest one, with the appropriate balance of cost, safety, and scientific, educational, industrial and economic ROI (Return On Investment). However, whatever else the architecture may be, it must be affordable. As was observed with the 90-Day Study, with its bottom line of $450 billion, an excessively high price tag simply makes the program nonviable.

By controlling costs while leveraging the latest technological and business innovations, combined with intelligent management, logistics and financing, the goal of sending humans to Mars is definitely within reach. Although further analysis is still necessary, with this architecture and an internationally collaborative approach it should be possible to implement a full program, including robotic precursor missions and three crewed surface missions, for well under $100 billion, and perhaps even less than half that amount.

For comparison, consider the following:

- In 1990, the estimated cost of the Space Exploration Initiative, which included a permanent base on the Moon and human missions to Mars, was calculated at $400-$500 billion over 20-30 years (the 90-Day Study).

- The International Space Station had cost an estimated $150 billion by 2010.

- NASA estimated the cost of "Mars Semi-Direct", the architecture developed for the original Design Reference Mission, at $55 billion over 10 years.

- Dr Robert Zubrin estimated that Mars Direct would cost $30-$50 billion over 20 years for five missions, if implemented by NASA; but only $5 billion if implemented by private enterprise.

- Mars One estimate that the cost of the getting the first crew of four people to Mars will be $6 billion, including precursor missions.

For the purpose of this exercise, the cost of the first three missions, including precursor missions and hardware development, is estimated at $36 billion. Although this is still very much a guess, it's based on NASA's estimate for the cost of the DRA, revised downwards to accommodate the cost reduction ideas incorporated into the plan.

This figure amounts to an average cost per mission of $12 billion, or $2 billion per crew slot. This is an average of $100 million per crew slot, per year, for two decades (~2020-40), which should be affordable by the top space-faring nations. Note, however, that the bulk of this expenditure will be required during the initial 20-year period of preparation and development, and securing financial commitment from all partners for this will be a challenge.

The intention is to run more than three missions. Consider that 75 missions have been run to the ISS so far. Because $150 billion was spent on constructing the space station, naturally the partners have sought to get their money's worth. The same logic applies to the IMRS, and the longer-term goal would be to invest on the order of $100 billion and run at least 10 missions. This will take about 33 years (2020-2053), during which time both architecture and base will evolve as a result of new discoveries and technological innovation.

The reason for making an estimate for first three missions, rather than 10, is for comparison with the DRA, of which Blue Dragon is an evolution.

The proper goal of a human Mars mission program should be sustained exploration followed by settlement. This can only be done if costs are kept low.
- Dr Robert Zubrin

1.3. International Partnership

It is the intention that the IMRS be developed as a true international collaboration, for the following reasons:

- To reduce the financial burden on any one nation.

- To open up Mars for all humanity rather than only one nation, race or ideology.

- To improve international relations by further developing trust, communication and cooperation between the world's major powers, thereby helping to bring about world peace.

- To take advantage of the people, facilities, hardware, technology, data, knowledge and other resources available in many countries instead of just one.

- To gain greater leverage from the research, expertise, designs and philosophies of the most talented scientists, engineers, managers and others from all over Earth, thus advancing space technology further and more quickly, reducing costs, benefiting Earth and all humanity, and increasing the overall value of the project's outcomes.

Sharing the cost

The IMRS plan is made yet more viable by distributing the cost across a consortium of multiple space agencies in a similar fashion to the ISS.

Because the economies of the participating nations are quite different, and because $100 million per year would represent a fairly significant fraction of some space agency budgets, it's possible that investment could be made partially or wholly "in kind" rather than "in cash". For example, a space agency may contribute an equivalent value in engineering, launch services, hardware or facilities.

It is hoped and expected that future missions will be progressively cheaper, and many other nations, perhaps even those that don't yet have a HSF (Human Space Flight) program, will eventually also be able to send astronauts to Mars.

1.3.1. Expanding the ISS Partnership

The ISS has been a huge success in terms of scientific research as well as international collaboration in outer space. The ISS partnership of five leading space agencies provides an excellent basis for expansion into a larger partnership for development of the IMRS.

The five ISS partners are:

1. NASA

2. Roscosmos (Russian Federal Space Agency)

3. ESA (European Space Agency)

4. JAXA (Japan Aerospace Exploration Agency)

5. CSA (Canadian Space Agency)

These countries are represented by flags in the upper half of the IMRS emblem (see The IMRS Emblem).

The proposed new partners are:

1. AEB (Agência Espacial Brasileira/Brazilian Space Agency)

2. ISRO (Indian Space Research Organisation)

3. KARI (Korea Aerospace Research Institute)

4. CNSA (China National Space Agency)

5. SSAU (State Space Agency of Ukraine)

These countries are represented by flags in the lower half of the IMRS emblem.

Brazil is a bilateral partner with NASA in the ISS, developing flight equipment and payloads in return for access to ISS facilities and flight opportunities for Brazilian astronauts. With this established relationship with the ISS, they should be included in the IMRS program.

The space agencies of South Korea and India expressed their desire to join the ISS in 2009, and talks commenced in 2010. Although this goal has not yet been realised, these are nonetheless leading agencies and should be included in the IMRS partnership. Both nations have their own astronauts and rockets, and India recently became the first Asian nation to reach Mars orbit.

China operate their own independent space station (Tiangong), and, along with the US and Russia, are one of only three nations with human spaceflight and lunar soft-landing capability. They are currently building one of the largest rockets ever conceived — the Long March 9 — and have plans for human missions to the Moon and Mars. They are one of the world's leading space agencies and should definitely be included in the IMRS.

When interviewed in 2010, then ESA Director-General Jean-Jacques Dordain said his agency was ready to propose to NASA and the other space station partners that China, India and South Korea be invited to join the station partnership, saying: "These three nations have been active in the multilateral discussion of future space exploration architecture. It seems that these three would be a good place to start widening the partnership." (Selding 2010).

Ukraine have their own astronauts and launch vehicles, and plans for lunar exploration and development of advanced space technology. SSAU are one of the top 10 government space agencies in terms of funding (excluding European agencies other than ESA), and are currently engaged in a joint venture with AEB to develop new launch facilities in Brazil. They are also a member of ISECG (International Space Exploration Coordinating Group — see below). It will be highly advantageous to include Ukraine in the IMRS partnership.

The International Space Exploration Coordinating Group

In 2007, as a response to a report titled "The Global Exploration Strategy — The Framework for Coordination," (Global Exploration Strategy 2007) the ISECG was formed from a consortium of 14 space agencies:

1. ASI (Agenzia Spaziale Italiana/Italian Space Agency)

2. CNES (Centre national d'études spatiales/National Centre for Space Studies) (France)

3. CNSA

4. CSA

5. CSIRO (Commonwealth Scientific and Industrial Research Organisation) (Australia)

6. DLR (Deutsches Zentrum für Luft- und Raumfahrt/German Aerospace Center)

7. ESA

8. ISRO

9. JAXA

10. KARI

11. NASA

12. SSAU

13. Roscosmos

14. UKSA (United Kingdom Space Agency)

Assuming that the space agencies of Italy, France, Germany and the UK will be represented by ESA in the IMRS partnership (as they are in the ISS), there's almost a total overlap between this list and the 10 space agencies above, suggesting that the IMRS could potentially be developed in conjunction with ISECG.

AEB is not yet part of ISECG, although considering its history of collaboration with NASA and SSAU, and involvement with the ISS, perhaps it should be. CSIRO is not listed as an IMRS partner mainly because it doesn't have a human spaceflight program; however, CSIRO could certainly participate in space science and engineering aspects of the IMRS. (Naturally it's the author's hope that Australia will revise its feeble space policy and establish a well-funded Australian Space Agency actively involved in human spaceflight and interplanetary exploration, giving it the credibility to be a full IMRS partner. One can dream.)

Several nations can afford to send their own robotic exploration missions to Mars but there are significant benefits in coordinating these national efforts and future human exploration missions.
- Global Exploration Strategy

Space is for peace

At this time in history it may be difficult to imagine some of the IMRS nations collaborating in space; or, for that matter, anything else. However, in addition to the myriad other benefits of space exploration, collaboration in space has proven an effective peacemaker. The US and USSR competed in the space arena for 20 years before commencing a long period of cooperation, to the point where, for a period, the US relied on Russian spacecraft to ferry its astronauts to and from the ISS. In fact, the two nations were cooperating in space long before the end of the Cold War in 1991. Three decades earlier, in 1961, John F Kennedy said "Let both sides seek to invoke the wonders of science instead of its terrors. Together let us explore the stars." Talks began about US-Soviet collaboration in space, and in 1975, the Apollo-Soyuz Test Program, which developed compatible docking systems, signified the beginning of a long period of collaboration between the two superpowers.

Perhaps because activities in space take place above the surface of the Earth, they also seems to be above its petty squabbles. Space is inspiring and transcendent, and virtually all who work in the field have a tacit understanding that expansion into space represents a solution to many human problems. Astronauts, scientists and others involved in space exploration seem to be aware, at least at some level, that they are working to create a future in which international conflict has ended and humanity expands into the cosmos as a united family.

Although there may be geopolitical tensions currently present on Earth, by setting the intention to create a truly international Mars program and establishing a framework for collaboration on Mars settlement for the benefit of all humanity, those tensions can be eased and ultimately eliminated. The result will be a foundation for greater global unity and a positive, abundant human future.

Astronaut Ron Garan expresses this sentiment well:

> I found myself looking down at the enormous International Space Station, one hundred feet below me, against the backdrop of our indescribably beautiful planet 240 miles below. The sheer beauty of the scene took my breath away— but even more compelling than the beauty was the realization of the tremendous human achievement that the International Space Station represents. It is not only an amazing technical accomplishment— probably the most complex structure ever constructed— but also one of the most amazing examples of international cooperation. As I hung there, looking down at the station against Earth, I marveled that fifteen nations, some that have not always been the best of friends, had found a way to set aside their differences and achieve something amazing in space. I wondered what the world would be like, and how many fewer problems we would all face, if we could figure out how to have the same level of cooperation and collaboration in our interactions on Earth's surface.

The purpose of the IMRS is therefore to achieve two great things at once; open up a new world for human civilisation while simultaneously bringing Earth to a condition of improved peace, abundance and environmental health. A key factor in achieving this is letting go of the "us and them" paradigm and realising that we're one global family, and that national borders are illusory and obsolete. This truth becomes abundantly clear when viewing Earth from space, and is known as the "overview effect".

Assertions are often made that various problems should be addressed on Earth before settlement of space. The truth is that expansion into space will greatly assist in solving virtually all of Earth's major problems, through inspiration, knowledge, understanding, cooperation, technological evolution, and access to unlimited resources.

Human expansion into space will help to bring the people of the world together, and should therefore be done together.

1.3.2. Ten Partners

If this ambitious collaboration can be achieved, the partnership will include 10 agencies. A model for financial participation in the first three missions could be organised into three tiers based on the level of the agencies' funding. This would determine the number of astronauts each can send within the first three missions:

| Level A
3 crew slots each	USA	NASA
	Russia	Roscosmos
Level B
2 crew slots each	Europe	ESA
	Japan	JAXA
	India	ISRO
	China	CNSA
Level C
1 crew slot each	Canada	CSA
	Brazil	AEB
	South Korea	KARI
	Ukraine	SSAU

At a cost of $100 million per crew slot per year for two decades, the total investment by "Level A" partners, with one astronaut on each mission, would

24

therefore be approximately $6 billion each. The total investment by "Level B" partners would be about $4 billion each, and the total investment by "Level C" partners would be about $2 billion each.

Crew slots

All international partners will be represented by at least one astronaut. This will give each partner a greater stake in the program; a greater return on investment in terms of national pride, recognition, and enthusiasm for STEM (Science, Technology, Engineering and Mathematics) education; and first-hand experience of a human interplanetary mission.

In order to preserve the truly international nature of each crew, and to permit a larger international partnership with each participating space agency represented by at least one astronaut, each could be restricted to a maximum of one astronaut per mission.

Below is a potential crew slot schedule for the first three missions. Alfa Mission favours those nations who currently operate space stations, namely the five ISS partners plus China, as these countries have the most experience with HSF.

	Crew slot 1	Crew slot 2	Crew slot 3	Crew slot 4	Crew slot 5	Crew slot 6
Alfa Mission	🇺🇸	⬛	🇪🇺	●	🇨🇦	🇨🇳
Bravo Mission	🇺🇸	⬛	🇪🇺	🇰🇷	🇮🇳	🇨🇳
Charlie Mission	🇺🇸	⬛	🇧🇷	●	🇮🇳	⬛

Level D partners

There are a number of other countries with space agencies that also have astronauts, including Bulgaria, Colombia, Indonesia, Israel, Malaysia, Mexico, Turkey and Vietnam. The goal should be to ultimately make flight opportunities to Mars available to all these nations, and indeed any others that commence astronaut programs in the meantime. With a minimum of 10 missions to Mars, slots could be found for astronauts from all these countries, which would generate a massive benefit for them, and for the world.

Any mission to Mars is likely to be a global effort.

- NASA Administrator Charles Bolden

1.3.3. The IMRS Emblem

The emblem for the IMRS depicts the planet Mars with 12 flags around it:

IMRS emblem

The 10 smaller flags represent the proposed partnership comprised of the world's top 10 space-faring nations (considering Europe as a single nation for the purpose of this exercise). This includes the five nations who created the ISS (USA, Russia, Europe, Japan and Canada), which appear at the top, plus five new partners (Brazil, China, India, South Korea and Ukraine), shown at the bottom. All of these nations are capable of participating in the IMRS and sending their own astronauts to Mars.

The large flag at the right of the emblem is the red-green-blue tricolour, which is the generally-accepted flag of Mars. This flag flies atop FMARS. It was designed by Pascal Lee in 1999 and, like the Mars trilogy by Kim Stanley Robinson (*Red Mars*, *Green Mars*, *Blue Mars*), is inspired by the colours of a changing Mars during human habitation. Mars is currently red, due to the ubiquitous presence of iron oxide across its surface. As terraforming proceeds, plants will grow on Mars, causing it to become green. And, eventually, when liquid water flows on the surface to form lakes and seas, and the atmosphere has become oxygenated, it will be blue like Earth.

The presence of the Mars flag in the emblem is intended to symbolise that the goal is not merely to visit Mars — i.e. this is not just a "flags and footprints" exercise — but to establish a new branch of human civilisation. This symbolism imbues the project with a much grander purpose that will help to inspire participants and supporters.

The large navy blue flag at the left of the emblem is the generally-accepted flag of Earth. It features the Blue Marble, a famous photograph of Earth taken by the crew of Apollo 17. Its role in the emblem is to symbolise that Mars is being explored, not by a single nation or even a group of nations, but as "the people of Earth", unified, for the benefit of all humanity, and for Earth herself.

2. Hardware

Blue Dragon has been inspired by the emergence of new space hardware that is both more advanced and cheaper than what has previously been available. Most of this is being developed by a new breed of space companies that are more agile and ambitious than their predecessors. This new style of space company, as well as the general shift in focus within the space industry from the public to private sector, is sometimes referred to as "NewSpace".

Although much of the required hardware is available from commercial vendors, some key components necessary for HMMs still need to be developed. However, in some cases they can be at least partly assembled or developed from existing available components.

2.1. Commercial Hardware

The most important goals of IMRS are to reduce costs and increase safety and the likelihood of success compared with existing proposals. These goals are all arguably made more achievable through utilisation of primarily COTS hardware.

The current era is different to any in which a HMM has been developed before. The private space sector is experiencing a revolution characterised by an exponential growth in entrepreneurial startups. This is somewhat related to, and reminiscent of, the IT sector during the past three decades. Numerous companies are now developing commercial space hardware and providing a wide range of space-related services, including launch services; space station and spacecraft components; satellites; space suits; ECLSS (Environment Control and Life Support System) hardware; space electronics; robotics; and other miscellaneous space products and services.

The cost of purchasing these items from commercial suppliers is much less than developing them from scratch. By taking advantage of this proliferation of commercial products and services, the cost of transporting people and equipment to Mars can be significantly lower than previous estimates. As the design for the IMRS evolves, the intention is to leverage new developments in COTS hardware where possible in order to further improve the architecture.

Rather than having a collection of one-off custom-built components that only a few specialist engineers understand, using COTS components gives several advantages:

- COTS components often have a greater operational maturity and a higher TRL (Technology Readiness Level) than custom-built hardware. They may have been used many times in multiple real-life applications, which has enabled refinement and improved understanding of the design. This drastically improves confidence in the technology and reduces the likelihood that unanticipated design flaws will manifest during use.

- COTS hardware is usually understood by a larger number of people, including engineers, customers and others who've used or studied it, especially if the product is popular or has been in use for some time. This makes problem identification and resolution quicker, easier and more likely to be correct.

- COTS hardware is usually cheaper than purpose-built hardware because with each successive production run optimisations are made in the product design, supply chain and manufacturing processes.

- The efficiency of mass production means that components produced in quantity can often be orders of magnitude cheaper than custom-built items.

It's always more expensive to build a prototype than a reproduction, because creating a prototype requires numerous iterations, modifications and redesigns, and several usually need to be made before the final version is considered ready for reproduction. When using COTS hardware, that work has already been completed and paid for, saving both time and money.

The total lifecycle cost of any hardware component in a space mission may be calculated using the following formula:

$$C_{total} = C_{dev} + C_{man} + C_{trans} + C_{op} + C_{dec}$$

where:

C_{total} = the total lifecycle cost of the item

C_{dev} = the cost of development

C_{man} = the cost of manufacture

C_{trans} = the cost of transportation and installation

C_{op} = the cost of operation

C_{dec} = the cost of decommissioning

In a traditional space program C_{dev} is a major expense, potentially on the order of billions of dollars, depending on the item. The main benefit of using COTS hardware is that this value is considerably lower per item because it's recovered across multiple sales, and shrinks towards zero over time.

In addition, C_{man} is often lower due to the cost benefits of mass production. Even C_{trans}, which is the total transportation cost from the manufacturer to the final destination (in this case, Mars), may be lower if a transport system designed especially for the item is available, such as is the case with the SpaceX Falcon rocket family and the Dragon capsule. C_{op} and C_{dec} may also have been optimised during the history of the component's usage.

COTS hardware suppliers sometimes collaborate in ways that support the assembly of mission architectures. For example, SpaceX and Bigelow Aerospace will both use the NDS (NASA Docking System), which will enable Dragon capsules to dock with Bigelow's inflatable space station modules for delivery of cargo and crew. This will be a valuable design feature that will reduce development costs.

Although a relatively new company, SpaceX is a fundamental player in the IMRS plan because their hardware is more advanced yet comparatively cheap. Focusing on a single technology (Dragon capsules) for crew transport to and from space, and using fundamentally the same tech to deliver cargo, reduces mission complexity and cost. Of course, when reusable rockets and capsules that can land on solid ground are developed by other companies or space agencies, these could be substituted in the architecture.

It's likely that major space agencies such as NASA, Roscosmos and CNSA will want their own hardware, particularly spacecraft, to be used in the missions, thereby justifying the investment made in development of those items and accentuating their importance. Although this could potentially inflate the overall cost of the program, it may be necessary for political reasons. Satisfying all partners will require successfully and diplomatically balancing their desires with practical cost and safety considerations, and it will be important for them all to understand that a decision to use hardware from any one supplier is not intended to deliberately favour that supplier's country, but simply to ensure that the program is affordable, viable and successful.

2.2. Rockets

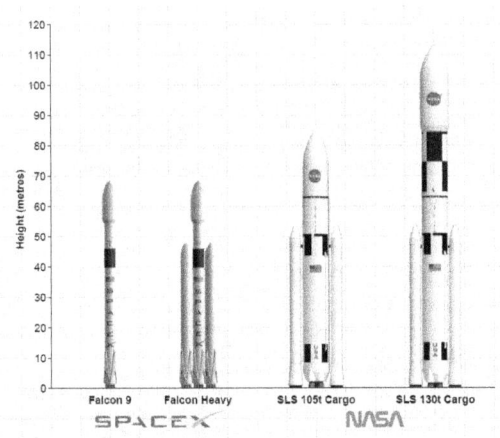

Launch vehicles for IMRS

2.2.1. SpaceX Falcon Rockets

Space Exploration Technologies Corporation, more commonly known as SpaceX, is one of the best known NewSpace companies. They're currently developing a new family of rockets called Falcon, powered by rocket engines of their own design called Merlin, which run on RP-1 (Rocket Propellant-1). SpaceX is also currently working on a new methalox engine named "Raptor", which will power the Falcon upper stages and become the main engines for a much larger vehicle: the Mars Colonial Transporter.

The Falcon 9 rocket (so named for its 9 engines) has flown 8 times successfully so far, and has an impressive schedule of launches in its upcoming manifest. One of the Falcon 9's most attractive features is its price, with a cost (as at January 2015) of only $61.2 million per launch. It's the first rocket to be designed and manufactured entirely in the 21st century.

Launch of a Falcon 9 (Credit: SpaceX)

The Falcon Heavy is currently in development and will start flying late 2014 or early 2015. It's essentially a Falcon 9 with two additional cores, for a total of 27 engines, and will be capable of delivering 53 tonnes of payload to LEO (Low Earth Orbit) or 13 tonnes to the surface of Mars. This will make it the most powerful rocket available, and the most powerful since the Saturn V, which carried astronauts to the Moon. The price of a Falcon Heavy (as at January 2015) is $85 million per launch.

SpaceX's modern processes and standards have produced launch vehicles that are significantly lower in price, yet more advanced than those of its competitors. Most importantly, SpaceX have begun making its rockets reusable. The recent "Grasshopper" test showed a reusable first stage ascending to a height of several hundred metres before smoothly descending and gently landing on its tail.

The "Grasshopper" Falcon first stage landing on its tail (Credit: SpaceX)

According to Elon Musk, CEO (Chief Executive Officer) of SpaceX, rapidly reusable rockets will reduce the cost per kilogram of sending material into space by a factor of approximately 100. Whether this is achievable or not remains to be seen, but it's obvious that reusability will have a significant effect on launch prices.

Thus, there are two main reasons why the SpaceX Falcon rockets are likely to be the best choice for this architecture:

- They're designed to carry Dragon capsules, which are fundamental to the architecture.

- They will be reusable, which will make them much cheaper.

SpaceX is also designing a much larger, super-heavy-lift vehicle called the Mars Colonial Transporter, which will be capable of delivering 100 tonnes to the surface of Mars. IMRS is not currently designed to take advantage of the MCT, as, at the time of writing, this is only a fairly recent development and few details are available.

Instead, the architecture employs the SLS (Space Launch System).

2.2.2. NASA Space Launch System

The SLS is a new family of SHLLVs currently in development at NASA, which has been developed within the following basic constraints:

- Capable of supporting human missions to Near Earth Asteroids, Earth-Moon Lagrange points, the Moon and Mars.

- Derived from tried-and-tested Shuttle components in order to reduce development costs and gain maximum leverage from existing infrastructure and material and human resources.

There are three planned configurations of the vehicle, referred to by the mass they can deliver to LEO: 70t (70 tonnes), 105t or 130t. The 105t and 130t versions support both crew and cargo modes.

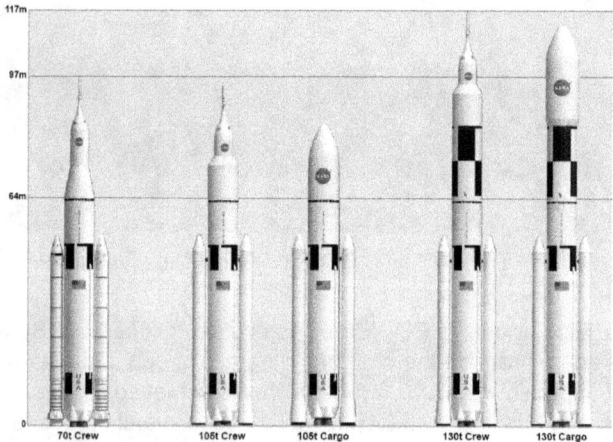

SLS vehicle configurations (Credit: NASA)

The SLS 130t Cargo will probably be the first launch vehicle capable of delivering 20-30 tonne payloads to Mars, making it suitable for delivering the SHAB and MAV to Mars.

Considerable work is currently being done on the SLS at NASA, and contractors such as Alliant Techsystems and Boeing have been engaged to develop components. Other SHLLVs are currently being studied at CNSA and Roscosmos, which may possibly be substituted in the architecture for the SLS if practical.

2.3. SpaceX Dragon Capsule

SpaceX have developed a space capsule known as Dragon, as part of NASA's Commercial Resupply Services program. Its primary purpose is for transporting cargo to and from the ISS. SpaceX made history in 2012 when the Dragon became the first commercial spacecraft ever to dock with the ISS.

Dragon capsules are designed to be carried by SpaceX's Falcon 9 and Heavy rockets. The current version of the Dragon is designed to splash down in water, like those used in NASA's Mercury, Gemini and Apollo programs and like NASA's new Orion capsule. However, the latest version of the Dragon capsule, known as Dragon V2, is capable of landing on solid ground.

Elon Musk with the Dragon V2 (Credit: SpaceX)

The Dragon V2 features eight SuperDraco engines, which are a powerful new variant of the Draco engines used by the Dragon's RCS (Reaction Control System). Like the Draco, the SuperDraco engines use non-cryogenic propellant, namely monomethyl hydrazine fuel and nitrogen tetroxide oxidiser. However, they're about 200 times as powerful, capable of delivering about 67 kN of axial thrust for a total of about 534 kN. The SuperDraco engine is the first ever rocket engine with 3D printed parts.

SpaceX SuperDraco engines (Credit: SpaceX)

The SuperDraco engines enable the Dragon V2 to land propulsively on solid ground, thus saving the time and expense of water recovery. On Earth the capsules will usually land at the original launch pad, but they can also land on the Moon, Mars, or other worlds with solid surfaces. This is in alignment with Musk's vision of establishing settlements on Mars.

The Dragon V2 can accommodate up to seven astronauts, and will be used for transporting crew between Earth and the ISS in the near future, as part of the NASA Commercial Crew Program.

In Blue Dragon, which is designed for a crew of six, the intention is to replace the lower centre seat with a container for cargo. On the way to Mars, this may include last minute supplies; on the way back, samples from Mars. All three crew Dragons in the architecture will be modified in this way to accommodate six people plus storage.

Interior of the Dragon V2 (Credit: SpaceX)

Dragon V2 control panel (Credit: SpaceX)

Red Dragons

NASA has begun researching a Mars lander called Red Dragon, which is a proposed variant of the SpaceX Dragon capsule that will provide a comparatively low cost technology for delivering payloads to the surface of Mars (Karcz et al. 2012).

Red Dragon landing on Mars (Credit: SpaceX)

Red Dragon will also be configured with SuperDraco engines, which are powerful enough for landing propulsively on Mars. In addition, Red Dragon will incorporate several modifications necessary for EDL on Mars, including:

- Removal of systems unique to LEO missions, such as berthing hardware.

- Addition of deep space communications.

- Modifications to suit the Martian environment.

- Algorithms and avionics for pinpoint landing on Mars.

The gravity on Mars is much lower than on Earth, which means the acceleration of the capsule towards Mars is comparatively lower. However, in the case of direct entry, the capsule will be approaching from interplanetary space at a much higher velocity than if it were descending to the surface of Earth from Earth orbit. Mars's atmosphere is much thinner (less than 1%) than Earth's, and will play less of a role in reducing spacecraft velocity during EDL; for the same reason, the effect of atmospheric friction will be less. These different conditions will affect the forces experienced by the spacecraft during EDL, which will necessitate changes to thrusters, heat shield, avionics and other parts of the spacecraft.

A Red Dragon capsule is estimated to be capable of delivering payloads of up to about 1.9 metric tonnes to the surface of Mars. This delivery mechanism has been receiving increasing attention from NASA, being considerably simpler and cheaper than existing landing techniques; for example, the sky crane method used to deliver the Curiosity rover. Not only will it be cheaper per kilogram of payload mass, but cheaper overall.

With human missions, another advantage is that a landed Dragon capsule can be repurposed as a storage unit, laboratory, shelter or mini-habitat.

Once the Red Dragon technology has been proven as a reliable mechanism for delivery of cargo, this approach may be used to deliver up to seven crew members to Mars surface by using a Dragon V2 modified in the same way.

Red Dragons represent near term technology that can enable comparatively inexpensive and functional Mars missions. They are fundamental to Blue Dragon.

The Dragon V2 capsules are being designed to land with a high degree of accuracy. From the SpaceX website:

> *SuperDraco engines will power a revolutionary launch escape system that will make Dragon the safest spacecraft in history and enable it to land propulsively on Earth or another planet with pinpoint accuracy.*

This ability to land "with pinpoint accuracy" is provided by the Dragon's GNC (Guidance, Navigation and Control) system. Due to the lack of GPS (Global Positioning System) on Mars, high-accuracy landings must be achieved using alternate methods. This problem has effectively been solved. For example, ESA have been developing a system known as LION (Landing with Inertial and Optical Navigation) that will enable pinpoint landing on the Moon, Mars and asteroids using image recognition of major landmarks (Delaune et al. 2012).

Another important development is the Fuel Optimal Large Divert Guidance (G-FOLD) algorithm (Acikmese, Casoliva & Carson 2012), able to autonomously calculate landing trajectories in real-time. This was recently tested successfully with Masten Space System's Xombie VTOL (Vertical Take-Off and Landing) experimental rocket, with the vehicle making a 750 metre course correction in real time.

Using these or similar technologies, the Red Dragons will be capable of pinpoint landings on Mars, and, because the position of landed base components can be known with precision, a neat, safe and optimised layout of the base can be planned.

These capsules potentially represent a mechanism for delivering cargo or crew to the surface of Mars that is not only repeatable, but affordable. It may soon be possible to deliver a payload to Mars for under $200 million, which is extremely cheap in terms of space missions.

Ice Dragon

NASA commenced studies of a mission to Mars based on the Red Dragon landing system, named "Ice Dragon" (Stoker et al. 2012). It was being developed in collaboration with SpaceX, and would have delivered a science package to Mars including a drill to penetrate up to two metres into the permafrost to investigate environmental conditions suitable for past or extant life. Despite being a highly valuable, practical and affordable mission, Ice Dragon was rejected by

NASA in favour of the Mars 2020 Rover.

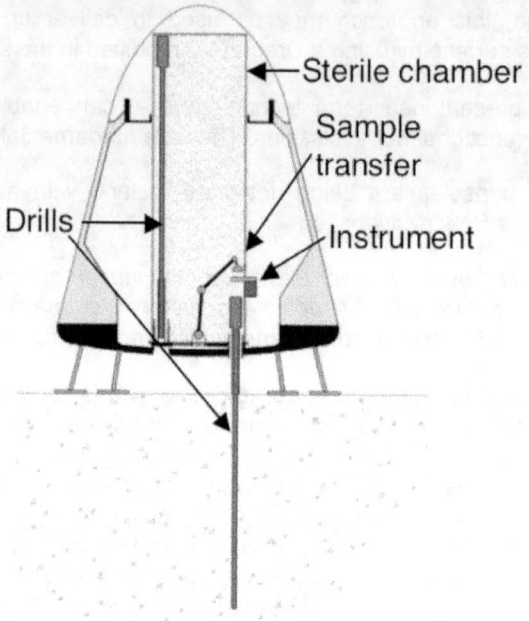

Ice Dragon mission (Credit: NASA)

The Ice Dragon mission was designed to address three main questions:

1. Is there life on Mars?

2. Are there viable and accessible resources for humans?

3. Is it safe to land humans on Mars?

The mission proposed to answer questions 1 and 2 by studying subsurface ground ice, and question 3 by testing the Red Dragon landing system. The value of investigating subsurface ice is described as follows:

> The subsurface environment provides protection from radiation to shield organic and biologic compounds from destruction. The ice-rich substrate is also ideal for preserving organic and biologic molecules and provides a source of H_2O for any biologic activity. Examination of martian ground ice can test the hypotheses of whether ground ice supports habitable conditions, that ground ice can preserve and accumulate organic compounds, and that ice contains biomolecules that show past or present biological activity on Mars. Furthermore, water on Mars, in the form of ground ice and hydrated minerals may provide a valuable resource to enable long-term human exploration. Water can provide the raw materials for rocket propellant, other chemicals and materials, and life support consumables for future human

Mars missions.

Besides the scientific outcomes of the mission, which would have been of tremendous value to human missions, one of the most important contributions of Ice Dragon would have been demonstration of the EDL capabilities of the Red Dragon capsule.

Larger capsules

Dragon capsules have a diameter of 3.7 metres. However, the architecture for the Mars One mission, which proposes to send 24-40 astronauts on a one-way mission to Mars, proposes to rely on a larger, 5-metre-diameter variant of the Dragon capsule for habitat modules. Although these are yet be to be built or demonstrated, Mars One's plan is to land the first two of these on Mars in 2020, which is only 7 years from the time of writing.

Mars One habitat capsules (Credit: Mars One, Bryan Versteeg)

SpaceX and Mars One do not have a formal association, and SpaceX have not announced the development of larger Dragon capsules. Therefore, Blue Dragon does not presently include them.

2.4. Bigelow Aerospace B330

The B330 module from Bigelow Aerospace is an inflatable space habitat slated to become available from about 2017. The B330s are designed for use in space stations, interplanetary vehicles, or surface habitats on the Moon or Mars, and include an ECLSS designed to safely support a crew of six long-term.

The structure of the B330 is essentially a solid central core surrounded by an inflatable, thick-walled "bubble". The core contains the ECLSS, bathroom, and possibly also food storage, air tanks and a solar storm shelter. At each end of the central core is a docking port and airlock. One end of the module has two large solar panel arrays and two thermal radiators.

B330 interior showing the central core (Credit: Bigelow Aerospace)

Aside from the basic dimensions and performance characteristics, few other details about the B330 are currently available.

The B330 modules may be launched in a deflated state, thus permitting them to be launched within a smaller diameter payload fairing, reducing launch costs. They may be connected together to form larger structures; for example, below is shown Bigelow Aerospace's concept for their "Alpha Station", which shows four connected B330 modules with three SpaceX Dragon capsules attached:

Alpha Space Station concept (Credit: Bigelow Aerospace)

B330 modules will use the NDS, which has been developed for the Orion Crew Vehicle and Commercial Crew Vehicles. SpaceX's Dragon capsules also use the NDS, and can therefore dock with B330 modules.

Bigelow Aerospace and SpaceX have already agreed to cooperate on interoperability between Dragon capsules and B330 modules.

The skin of a B330 is comprised of 24-36 layers for ballistic, thermal and radiation protection. The B330's radiation protection is approximately equal to that of the ISS, however, its ballistic protection is superior.

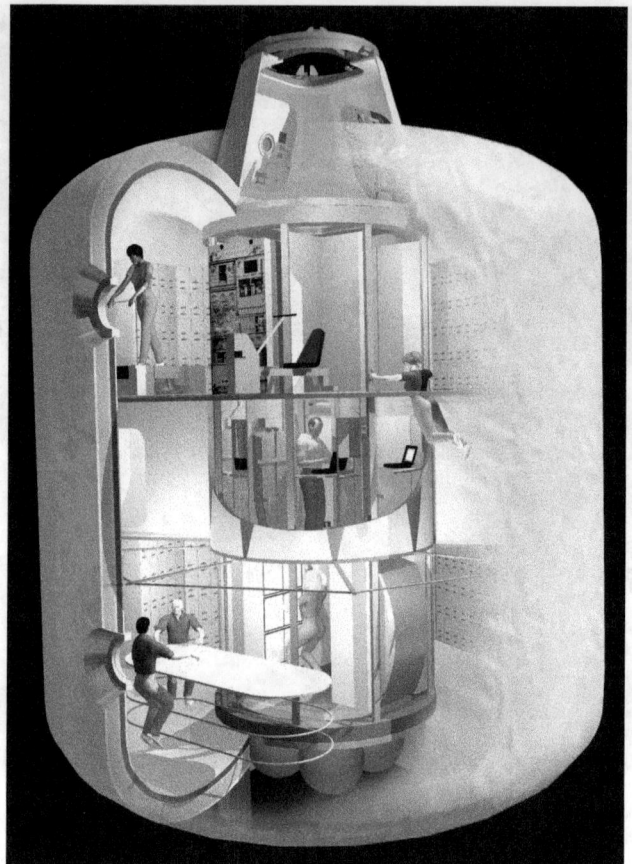

B330 cutaway (Credit: Bigelow Aerospace)

Internal Volume

The B330 modules have an internal volume of approximately 330 m³ (hence the name), which is ample for a crew of six, including storage for consumables. The bare minimum free volume required per person is generally considered to be 10 m³, although the preferred minimum is considered to be about 19 m³. In the B330, even if half the available volume is devoted to supplies and equipment there will still be 27.5 m³ free volume per person, which is quite spacious compared with previous crewed spacecraft and space stations.

B330s are comparatively much less dense than ISS modules and past space habitats. A B330 weighs about 20 tonnes, yet has approximately three times the internal volume of the ISS Destiny module, which weighs about 15 tonnes. This benefit, in addition to the fact that they can be launched in a deflated state, means that the B330 modules represent a greatly reduced cost per cubic metre of space habitat.

Versions

The base model B330 is designed for Earth orbit.

B330 on orbit (Credit: Bigelow Aerospace)

Bigelow Aerospace have said they will also provide a "Deep Space" version of the B330 optimised for interplanetary space, in additions to versions optimised for the surfaces of the Moon and Mars. The Deep Space version provides water storage in the form of "water tiles", which are thick, square containers of water, lining the interior of the habitat. As water is one of the most effective substances for absorbing radiation, this provides additional protection for the crew (Cohen, Flynn & Matossian 2012).

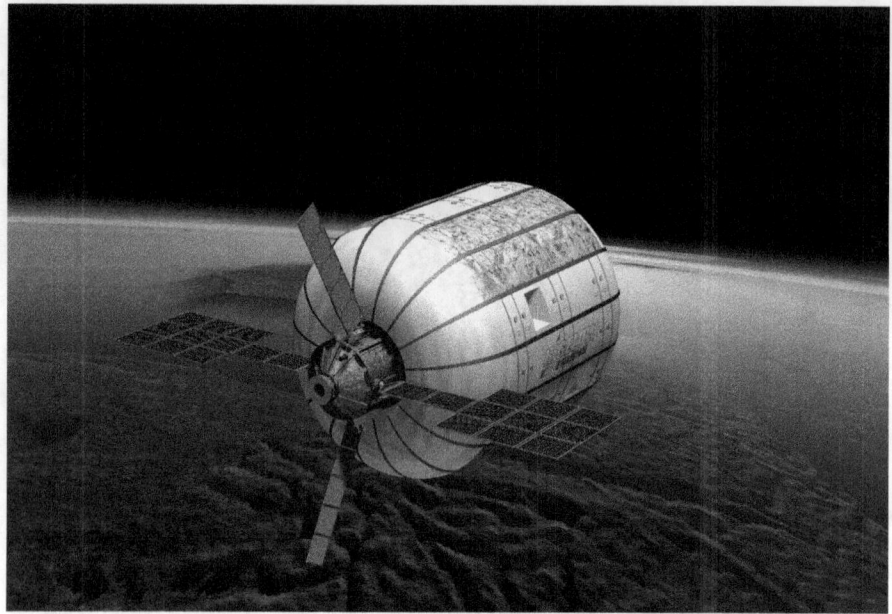

B330 on Mars orbit (Credit: Bigelow Aerospace)

Launching the B330

A B330 has a mass of approximately 20-23 tonnes, length of 14 metres, an inflated diameter of 6.7 metres, and an estimated deflated diameter of about 3.5 metres. The rocket currently under consideration by Bigelow Aerospace for launching these to orbit is the Atlas V. However, the only Atlas V configuration that could launch a B330 is probably the Heavy (HLV DEC/5H2), which can deliver 29.4 tonnes to LEO. As yet the Atlas V Heavy is yet to be built or flown, and it's estimated that preparing one would take about 30 months.

The Atlas V Heavy has a payload diameter of either 5.4 m and a payload length of 16 m, which is sufficiently large to launch a deflated B330. However, a Heavy may not be required, as the Atlas V is a flexible vehicle with a range of payload configurations, including diameters of 4 m or 5.4 m, and lengths ranging from 9 m to 16 m or even more, and it's possible that an alternative configuration may be developed specifically to suit the B330.

Another option could be the SpaceX Falcon Heavy. However, its 11.4 metre payload fairing would need to be extended, and the B330 would need to be launched in a deflated state to fit within the 4.6 metre diameter fairing.

Two B330 modules are required by the architecture: the THAB (Transit HABitat) and the SHAB. Although one of the main advantages of inflatable habitats is that they can be launched by smaller rockets in a deflated state, it may be preferable for this architecture to launch both habitats inflated and already fitted out, as this will enable many systems to be tested on Earth where engineers are available to make repairs and enhancements. It will enable training missions to be conducted on the ground, and the crew to become familiar with the equipment and the interior of their actual habitats (rather than mockups or prototypes) before the mission.

If the THAB is launched deflated, it will be necessary for one or more fit-out crews to visit it via Dragon capsule, which will be risky, expensive, time-consuming, and inconvenient. If the SHAB is sent to Mars deflated, it will not be in a habitable state when the crew arrive. After having spent 6 months in microgravity, it will not be practical for them to immediately have to begin assembling furniture and setting up equipment. It will be much better if everything is ready for them on arrival. This will also enable more of the habitats' systems to be remotely tested before the crew leave Earth.

With these things in mind, and considering the critical importance of the habitats to the mission, the additional cost associated with launching the habitats in an inflated state and fitted out is justified.

With its payload fairing of 8.4 m, the only vehicle likely to be available in the near future capable of launching a fully inflated B330 is the SLS. For launching the THAB the most suitable variant may be the SLS 105t Cargo. Its payload capacity of 105 tonnes is more than quadruple the mass of an empty B330, which means the THAB can be launched fully stocked with supplies, together with a section of *Adeona* (see <u>On-Orbit Assembly</u>).

An SLS 130t Cargo is necessary for delivering the SHAB to Mars, which should be able to deliver in the range of 20-30 tonnes to the surface of Mars.

2.5. Hardware Development

Components unavailable from commercial suppliers must still be developed, and this will probably require a substantial investment. However, current trends indicate this cost will decrease as more commercial space hardware suppliers and components become available, and as the space industry becomes increasingly confident in its own evolution and the expansion of the market. Technologies such as 3D printing and nanostructured materials are also driving down the cost of manufacturing, while simultaneously enabling superior designs.

Considering the vast range of technical services now available in the global space community, it should be possible to develop any remaining requirements within more modest time frames and budgets than have previously been considered. The development of any major component should begin with a comprehensive survey of products and services available from commercial suppliers.

Examples of technological components that still need to be developed include:

- A reliable EDL system for delivering 20-30 tonne payloads to the surface of Mars.

- A reliable vehicle for transporting a crew from Mars surface to Mars orbit (i.e. a Mars Ascent Vehicle).

- Proven technology for producing propellant, breathable air and potable water from the local Martian environment.

- Effective techniques and technologies for dealing with radiation and Martian dust.

The expense of developing these components is not necessarily a sunk cost. There's a clear desire in the world population for settlement of Mars, and initiatives such as Inspiration Mars, Mars One and IMRS will surely be followed up by other human exploration and settlement programs, both public and private. As these technologies will be of value to any human mission, they could potentially be licensed, sold, or developed into new COTS products, providing opportunities to recover the initial investment and make a profit.

3. The Architecture

The IMRS plan includes a new humans-to-Mars mission architecture that exploits some of the latest developments in space technology such as reusable rockets, capsules that can land on solid ground, inflatable space habitats and mechanical counter-pressure spacesuits. It is modern, practical, comparatively safer than its predecessors, and affordable. Although designed for multiple missions to a single location, which is the plan for IMRS, it would also be practical for missions to discrete locations.

In this document the architecture is described mostly in relation to high-level design choices without going into deep engineering detail.

The name of the architecture, "Blue Dragon," was inspired by "Red Dragon", the Mars landing system adapted from SpaceX Dragon capsules, as these capsules, which have the potential to safely and accurately land cargo and crews on Mars, provide the fundamental point of difference that enables improvements over the DRA. The architecture connects Mars and Earth; Mars is red, Earth is blue. Also, the dragon in the SpaceX Dragon logo is blue.

Blue Dragon is an evolution of the NASA Human Exploration of Mars - Design Reference Architecture (DRA) 5.0 (ed. Drake 2009). The DRA is arguably the most mature Mars mission architecture currently available, having been iteratively developed by hundreds of qualified people since 1995. Nonetheless, it has the potential to be further improved on the basis of new science, technology and ideas. The changes presented here improve safety, reduce cost, and achieve a superior overall result.

The IMRS program does not have precisely the same goals as the DRA, although they are very similar.

The DRA proposes a series of three more-or-less identical missions to Mars, each to different, scientifically interesting locations, and at these locations to perform exploration and scientific research of Mars; for example, analysing rocks, dust, subsurface ice and the Martian atmosphere, searching for past or extant life, and so on. In addition the DRA program proposes to research ISRU technologies, including production of oxygen, water and food, and strategies for long-term habitation of Mars.

The DRA is like "Apollo on Mars", in the sense that each mission in the Apollo program went to a different location on the Moon. Although the scientific research performed at the Apollo sites has been useful for investigation into future settlement of the Moon, none of the Apollo sites were considered as locations of future settlements, because settlement was not the main focus. Although it was hoped by many that Apollo would continue into the Apollo Applications Program and that a permanent human presence would be established on the Moon, for various reasons this never eventuated, and at that time the private sector was not in a position to undertake such a challenge itself.

In the 45 years since Apollo 11 there has been considerable ongoing interest in establishing a lunar settlement, with the Artemis Society and their associated public outreach group the Moon Society being major proponents, along with other enthusiast groups and major space agencies. However, as robotic exploration of Mars has progressed and understanding of the red planet has developed, it's apparent that a much stronger desire has emerged in the space community for settlement of Mars. Despite the emergence of companies such as Moon Express and Golden Spike, private missions such as Inspiration Mars and Mars One are receiving much more attention, as are the Mars settlement plans of companies like SpaceX and Virgin Galactic.

If membership in space enthusiast groups and public interest in missions are any kind of indicators, the number of people supportive of colonisation of Mars far outweighs that of Earth orbit or the Moon. Because of Mars's similarities with Earth, such as a day length of an approximately 24 hours, seasons, weather, a transparent atmosphere, a coloured sky, and familiar-looking terrain highly reminiscent of Earthian deserts, settlement of Mars is far more amenable to the human imagination and therefore has greater appeal. Its abundance of all the elements necessary for life and technological civilisation, including a range of options for energy production, make it a far more attractive and accessible settlement target than any alternative.

As humanity's technical capability develops, interest in settlement of Mars grows steadily. The facts that over 200,000 people applied for a one-way trip to Mars as part of the Mars One initiative, plus the consistent turnout to various international Mars conferences by enthusiasts and industry leaders, are strong indicators of this trend.

Inspiration Mars and Mars One represent the first two of what is likely to be an ongoing series of proposed missions to Mars initiated by private enterprise. These missions have developed from a growing interest in settlement of Mars, and have stimulated yet more interest. More missions will surely follow, both public and private, stimulated by advancements in computing, robotics, materials, space hardware, other branches of science, technology and engineering, and the ongoing evolution of the space economy.

3.1. Primary Components

These are the main hardware items utilised by the architecture.

Type	Component	Description
Launch vehicles	SpaceX Falcon 9	Medium lift launch vehicle for launching Dragon capsules and satellites to space.
	SpaceX Falcon Heavy	Heavy lift launch vehicle for launching Dragon capsules and surface vehicles to Mars.
	Space Launch System (SLS)	Family of heavy and super heavy lift launch vehicles currently in development at NASA, for launching the largest components.
Space vehicles	Mars Transfer Vehicle (MTV)	Spaceship that carries the crew to Mars and back. Comprised of the THAB plus propulsion stack. Also known as *Adeona*.
	Mars Ascent Vehicle (MAV)	Vehicle to transport the crew from Mars surface to Mars orbit. Includes ISPP system and *Kepler*.
Capsules	Mars Supply Capsule (MSC)	Cargo Dragon capsule modified for Mars landing (i.e. Red Dragon), for delivering supplies to the Martian surface.
	Earth-Mars Capsule (EMC)	Crew Dragon modified for Mars landing, which carries the crew from Earth surface to Earth orbit, then from Mars orbit to Mars surface. Also known as *Einstein*.
	Mars Ascent Capsule (MAC)	Modified Crew Dragon which forms the topmost section of the MAV. Also known as *Kepler*.
	Earth Descent Capsule (EDC)	Crew Dragon launched from Earth at the end of the mission, which the crew use to land on Earth. Also known as *Newton*.
Surface vehicles and robots	Crewed Adaptable Multipurpose Pressurised Exploration Rover (CAMPER)	A pressurised rover for multi-sol excursions on Mars, featuring robotic arms and attachments such as a drill and excavator bucket.
	All Terrain Vehicle (ATV)	Unpressurised 4-wheeled vehicle for 1-2 people.
	Autonomous Water Extraction from the Surface Of Mars (AWESOM)	Mobile robot that obtains water from permafrost around the base.
		Habitat where the crew live and work while

Habitats	Mars Transit Habitat (THAB)	*en route* to and from Mars. Part of the MTV and comprised of a B330 ("Deep Space" version).
	Mars Surface Habitat (SHAB)	Habitat where the crew live and work while on the surface of Mars. Comprised of B330 modified for Mars surface.
Satellites	Heliocentric Laser Communications Relay Satellite (HLCRS)	Laser communications relay satellite in heliocentric orbit.
	Mars Communications and Observation Satellite (MCOS)	Areosynchronous satellite for communication and observation.

Naming

The first mission to the IMRS is called "Alfa Mission" and the first crew is called "Alfa Crew"; the second mission is called "Bravo Mission" and the second crew is named "Bravo Crew"; and so on, following the International Radiotelephony Spelling Alphabet. This pattern has been adopted to support clear radio communications between different Mars exploration crews and Earth, which will become increasingly important when multiple missions are underway at the same time.

In order to mitigate acronym overload, friendly names, or at least, more memorable acronyms, are used for some of the components.

1. The MTV is a spaceship, and ships, generally speaking, should have a more interesting name than "MTV". The first MTV is therefore named "Adeona", the Roman goddess of safe return. The name comes from the Latin verb *adeo*, which means to approach, go to, visit, attend, undertake or undergo, as well as to take possession of one's inheritance. This is appropriate, as Mars is humanity's inheritance, and a safe return is desired for the crew.

2. The crew capsules are named after famous scientists. The name of the specialised Dragon capsule that carries the crew from the surface of Earth to Earth orbit, then from Mars orbit to the surface of Mars, is called "Einstein". This name came from Earth-Mars Capsule, or EMC, which is reminiscent of Albert Einstein's famous equation $e=mc^2$. The Mars Ascent Capsule is named "Kepler" after Johannes Kepler, due to his associated with Mars and orbital mechanics, and the Earth Descent Capsule is named "Newton" after Isaac Newton, due to his associated with gravity. (All of these were male Europeans, but please don't be offended if you aren't male or European — the names are just for the purpose of this

document!)

3. The pressurised rover is known as the Crewed Adaptable Multipurpose Pressurised Exploration Rover, or "CAMPER", which alludes to the vehicle's form and function. The robot that collects water from the Martian surface is called "AWESOM", which stands for "Autonomous Water Extraction from the Surface Of Mars".

Spacecraft names are shown in italics, as per the convention for aquatic ships.

3.2. Mission Overview

The following schedule gives an overview of the proposed first human mission to Mars. Subsequent missions can utilise substantially the same architecture, except predeploying different assets depending on the needs of the base.

Phase 1: Prepare (now-2030)

1. Improve TRL of the Red Dragon technology, marssuits, B330 habitat modules, and other hardware elements.

2. Develop and construct the SHAB, THAB, MAV, CAMPER, satellites, methalox rocket engines, propellant pods and other major hardware components.

3. Select location for the IMRS.

4. Select and train Alfa Crew.

5. Conduct precursor missions.

Phase 2: Predeployment (2031)

As the intention is to send the crew in 2033, the predeployment phase is scheduled for the launch opportunity prior, in 2031.

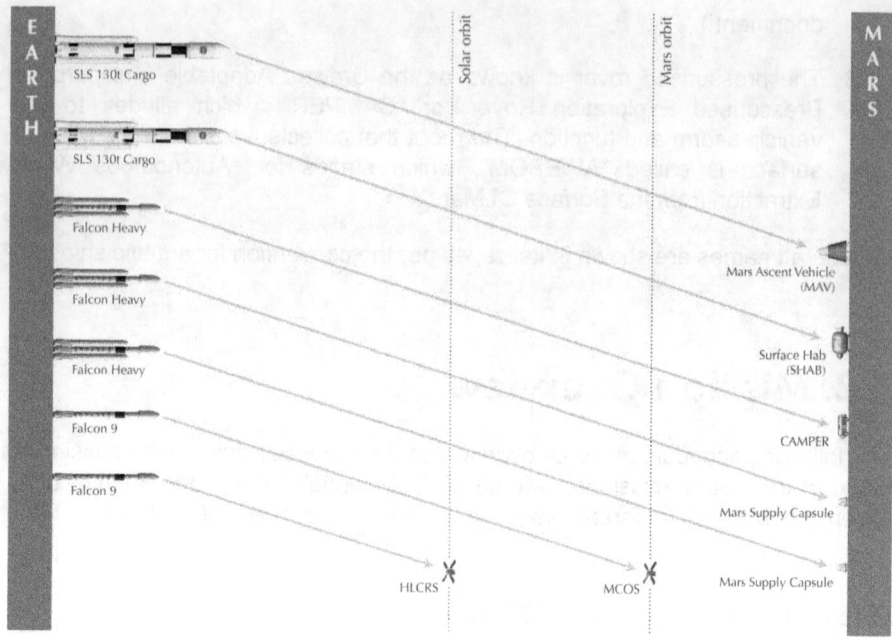

Bat chart - Predeployment

1. Launch the MAV, SHAB and CAMPER, land them safely at IMRS, remotely activate and test. Commence ISRU operations.

2. Pack and launch the Mars Supply Capsules, and send to the IMRS.

3. Launch first MCOS and place in areostationary orbit above IMRS.

4. Launch first HLCRS and place in heliocentric orbit.

5. Use the CAMPER to explore the surrounding terrain and find a suitable LZ (Landing Zone) for *Einstein*.

Phase 3: Construct *Adeona* (~2032)

Although the construction of *Adeona* is scheduled here for 2032, it could be done at almost any time before the 2033 launch. The sooner *Adeona* is built, the sooner she can be used for training missions.

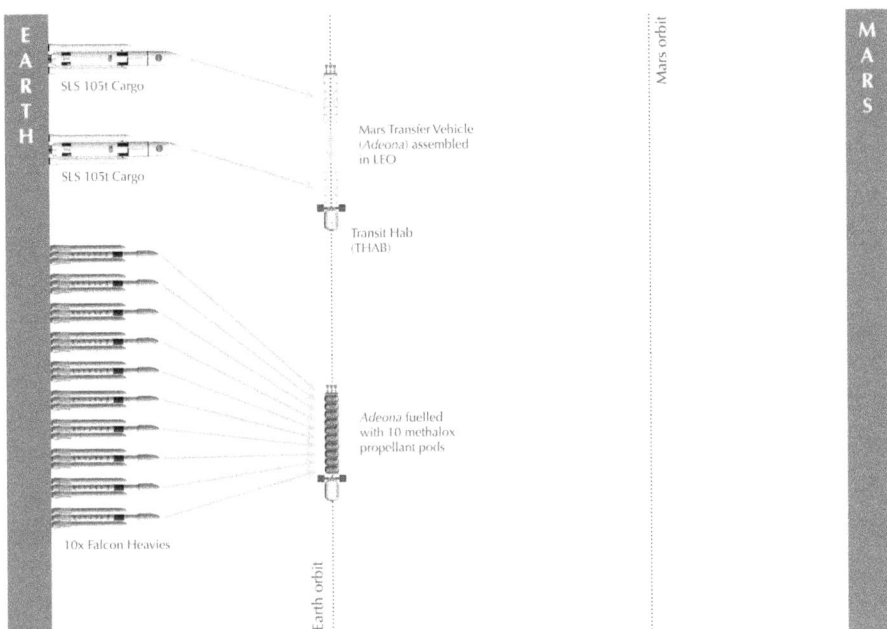

Bat chart - MTV construction

1. On Earth, inflate and fit out the THAB, and pack with supplies. Run training missions on the ground.

2. Construct the two halves of *Adeona*, and launch to LEO using SLS 105t Cargo rockets.

3. Assemble *Adeona* on orbit by connecting the two halves together.

4. Run training missions to *Adeona*.

5. Launch 10 reusable Falcon Heavies with full methalox propellant pods to *Adeona*.

Phase 4: Crew Outbound (2033)

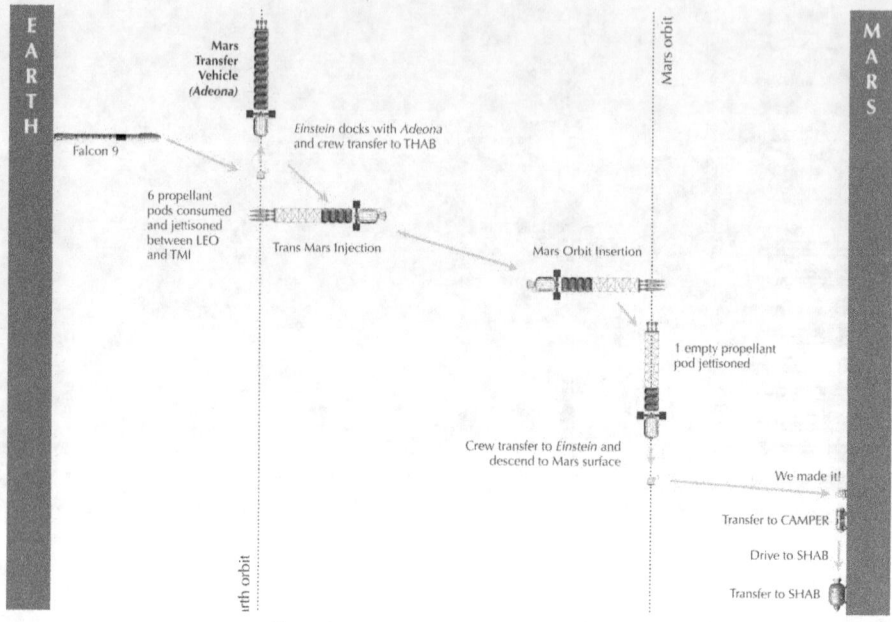

Bat chart - crew outbound phase

1. Launch the crew to LEO in *Einstein* atop a Falcon 9.

2. *Einstein* performs EOR (Earth Orbit Rendezvous) with *Adeona,* and crew transfers into the THAB.

3. With *Einstein* still docked, *Adeona* climbs from LEO to HEEO (Highly Elliptical Earth Orbit) in a series of six increasingly elliptical orbits, jettisoning one propellant pod after each burn to reduce mass.

4. *Adeona* performs the TMI (Trans-Mars Injection) burn and flies to Mars on a Hohmann MTO (Mars Transfer Orbit) for ~6 months.

5. Immediately prior to MOI, any unrecyclable waste generated during the outbound trip is jettisoned to reduce mass.

6. *Adeona* performs MOI (Mars Orbit Insertion) and parks on HEMO (Highly Elliptical Mars Orbit).

7. The 7th empty propellant pod is jettisoned to reduce mass.

8. The crew prebreathe and don their marssuits in preparation for Mars descent. *Adeona* is placed in standby mode.

9. The crew transfers from the THAB back into *Einstein*, and descend to the surface of Mars.

10. Under remote control by the crew, the CAMPER is driven from the SHAB

to *Einstein*'s LZ.

11. The crew exit *Einstein* and step onto the surface of Mars. They enter the CAMPER through its airlock, two at a time.

12. The crew drives the CAMPER to the SHAB.

13. The crew exit the CAMPER through its airlock, and enter the SHAB through its airlock, two at a time.

Phase 4: Surface Mission (2033-2035)

1. The crew spend 1-2 weeks in the SHAB, setting up their living space, testing systems, and adapting to Mars gravity and the SHAB's atmosphere.

2. The surface mission proceeds for ~18 months. During this time, the AWESOM robot, PV blanket and ISPP modules are moved from the MAV to the SHAB.

Phase 5: Crew Return (2035)

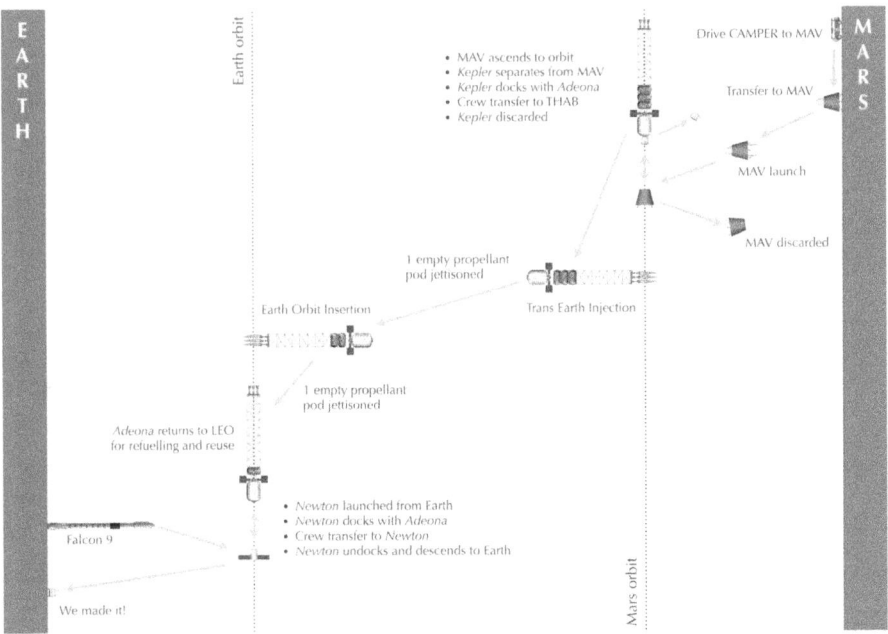

Bat chart - crew return phase

1. A full systems check is performed on the MAV and *Adeona*.

2. All samples for Earth return are stowed in *Kepler*.

3. The SHAB is cleaned and tidied, and placed in standby mode in preparation for arrival of the next crew in ~8 months.

4. The crew transfers to the CAMPER wearing marssuits, drive the CAMPER to the MAV, then transfer to *Kepler*.

5. The CAMPER is remotely driven back to the SHAB so it won't be damaged by the launch of the MAV.

6. The MAV, with crew, launches from Mars surface.

7. On approach to *Adeona*'s orbit, *Kepler* separates from the main body of the MAV, which is discarded. It will circle Mars for a while before falling back to the surface; or it could remain on orbit for future reuse.

8. *Kepler* performs MOR (Mars Orbit Rendezvous) with *Adeona*, and the crew transfers the samples and themselves to the THAB.

9. *Kepler* is undocked and left on Mars orbit for future recovery and reuse.

10. Any surplus contingency food and water is jettisoned to reduce mass.

11. *Adeona* performs TEI (Trans-Earth Injection) and flies back to Earth on a minimum-energy Hohmann MTO for ~6 months.

12. The 8th empty propellant pod is jettisoned to reduce mass.

13. *Adeona* performs EOI (Earth Orbit Insertion) to reach HEEO.

14. The 9th empty propellant pod is jettisoned to reduce mass.

15. *Adeona* performs an orbital transfer burn to return to LEO.

16. A Falcon 9 with *Newton* is launched from Earth.

17. *Newton* performs EOR with *Adeona*.

18. The crew transfers themselves and their samples from the THAB to *Newton*.

19. *Newton* undocks and performs EDL to Earth surface.

20. The crew and samples are quarantined.

Phase 6: Recover/Prepare

1. Alfa Crew is debriefed, and commence physical rehabilitation to restore fitness.

2. Charlie Crew is selected and trained. (Bravo Crew will have departed for Mars in the second MTV about 4 months prior to Alfa Crew's return.)

3. Missions are conducted to refurbish and refill *Adeona* with propellant in preparation for Charlie Mission in 22 months.

4. Additional hardware and supplies sent to the IMRS, such as solar panels, greenhouse modules, replacement parts, etc.

4. Key Features

In addition to using commercial hardware, the IMRS plan is based on a number of variations to the DRA intended to decrease cost, improve safety and increase the return on investment and the utility of the base. In this section several of the most important differences from the DRA, and their reasons, are discussed. These form the core design philosophies underlying the overall plan.

4.1. Predeploy the Hab; Land in Capsules

In Mars Direct, the crew travel to Mars, and land in, the habitat. In the DRA, the habitat waits on Mars orbit for arrival of the crew in the MTV, then they transfer to the habitat and use it for Mars descent. In Blue Dragon, however, the SHAB, like the MAV, is predeployed to Mars in advance of the crew, and activated and checked out remotely. Instead of landing the crew on Mars in a habitat, a Red Dragon capsule in crew configuration is used. There are several important reasons for this strategy.

Peace of mind

One of the innovations in Mars Direct is that the ERV (Earth Return Vehicle) is landed on Mars one launch opportunity (~26 months) prior to arrival of the crew. Once at Mars, the ERV's power system and propellant plant are activated and the methalox bipropellant is manufactured quickly to minimise hydrogen boil-off. In this way, mission planners know that there's an ascent vehicle on Mars, full of propellant and ready to go, before the crew even leave Earth. This idea greatly improves safety and was incorporated into the DRA; it is also included in Blue Dragon.

If predeployment is a good idea for the MAV then it applies equally well to the SHAB. The SHAB's systems can be tested and validated, including ISRU, power and communications, before the crew leave Earth. Knowing that the SHAB has safely landed and is fully operational will significantly improve confidence of mission planners and crew.

ISRU

Predeploying the SHAB creates a window of time during which breathable air and potable water for life support can be manufactured from local Martian resources. Taking this approach obviates the need to bring some or all of these resources from Earth, thus reducing mass and cost. Importantly, as mentioned below, this can be done without risk to the crew.

Safety

The SHAB will be heavier than anything else ever landed on Mars before, and because Earth's atmosphere and gravity are completely different to Mars's, it will not be possible to run a high-fidelity end-to-end test of SHAB EDL prior to its actual deployment. With the advanced computer modelling capability now available, a Mars landing can be simulated much more accurately than any aspects of space missions to date. Nonetheless, computer simulations are never 100% perfect. Some aspects of the SHAB's EDL system can be tested on Earth; for example, aerodynamic and aerothermal loads could be tested in Earth's upper atmosphere above 30 km altitude to simulate the thin Martian air. Terminal landing systems can be tested using a 0.38 mass scale model. However, although a high confidence in a successful landing of the SHAB can be achieved, the EDL system for delivering such a large and heavy item to the surface of Mars will not be properly tested until the actual Mars landing.

It will therefore be much safer to land the crew in a 6 tonne capsule rather than in a 20-30 tonne habitat or descent vehicle. This is especially true if the use of capsules for EDL has already been tested and proven to work several times. Before the first human mission, Dragon capsules will probably have been used many times for carrying crew and cargo between Earth orbit and surface, and possibly also several times for delivering experiments and cargo to Mars.

If the SHAB crashes on Mars there's no risk of LOC (Loss Of Crew), because they won't be in it. In that event, the design of the SHAB or its EDL system can simply be improved, and another built and sent, and this process can be repeated until success is achieved. A crew will only be sent to Mars once there's a fully operational MAV, SHAB, CAMPER, power system and everything else necessary for a successful surface mission already in place.

Cost

With a base-first strategy, if the architecture does not require landing in a habitat then each mission does not require another one; only another capsule. Since capsules will be significantly cheaper, this greatly reduces the cost of each mission.

4.1.1. Getting to the SHAB

Because the crew do not land in a habitat, a method for safely reaching the SHAB on arrival is necessary.

The capsule will not be able to land too close to the SHAB, or any other part of the base, because dust and debris thrown up by the engines during the landing may cause damage to the SHAB or cover the solar panels with dirt. The capsule LZ will therefore need to be at least a few hundred metres away from the SHAB, or at least downhill from it.

However, after spending 6 months in microgravity, it would be impractical to expect the crew to walk that far in marssuits immediately after landing on Mars, as they may have become weakened to some degree.

The solution is to pick the crew up in the CAMPER. After the dust has settled post-landing, the CAMPER will be driven from the SHAB to *Einstein*'s LZ under remote control by the crew. As a backup, MCC (Mission Control Centre) will also be able to remotely operate the CAMPER.

The path the CAMPER will take between SHAB and the LZ will have been driven along several times before they arrive. The LZ will be selected in advanced by exploring the surrounding terrain, both from orbit and by exploring the area with the CAMPER via remote control, and identifying a smooth and flat location with a clear path to the SHAB. Using the CAMPER's excavator attachment and/or robotic arms, the path can be made smoother, clearer and safer during the available time, thus improving the probability that the crew will reach the SHAB safely. This strategy relies on the ability to land *Einstein* with a high degree of precision.

The crew will be wearing MCP (Mechanical Counter-Pressure) marssuits during EDL, and will enter and exit the CAMPER and the SHAB through their respective airlocks.

If the CAMPER is unable to reach *Einstein* for some reason, then the backup plan will indeed be for the crew to walk. Although they may have become weakened during the outbound trip, if they have been diligent with their exercise and nutrition, as they should, then a short walk in the light Martian gravity should be achievable.

4.2. Multiple Missions

The DRA specifies a program of three missions for these reasons:

- A more substantial overall return on the necessary investment in hardware development, human resources and intellectual capital can be achieved if the architecture is implemented a minimum of three times.

- To complete the DRA three times will require a total of approximately one decade. During this period new technology will become available, the space industry will have evolved, and it will probably be appropriate timing to develop a more efficient or practical architecture.

This reasoning applies equally well to IMRS, and hence at least three missions should be planned. However, the intention is that this be the *minimum* number, and that at least 10 missions to the IMRS be conducted in order to establish some basic infrastructure, obtain scientific results, produce new inventions, and plant the seeds of settlement.

The goal is to establish a permanent human presence on Mars, a base for further exploration, and some fundamental infrastructure elements. Over the coming decades, as the technology develops, HMMs will become increasingly affordable and thus easier for for both government space agencies and private enterprise to conduct.

After each run-through of the architecture, opportunities to make improvements will present themselves and each mission will be progressively more efficient, faster, cheaper, safer, or otherwise an incremental improvement on the previous. The first mission will be the most expensive, after which the cost of each mission will decrease as the architecture, designs and hardware components are re-used and improved. Alternately, the same budget can be used to achieve more.

Re-use of the architecture multiple times produces an even greater value benefit in the IMRS than it does in DRA, because the missions are all targeted at the same location, therefore, much of the hardware delivered in the initial predeployment can be re-used by each successive mission. If an ongoing financial commitment can be secured from each participating space agency, this will enable the subsequent development of an updated architecture based on new technology and scientific information, and further exploration and development of Mars can continue.

4.3. Base-First Strategy

Once the IMRS has been built, numerous missions can be conducted to the same location at comparatively low cost. By locating the IMRS in an interesting area and including a long-range surface vehicle such as the CAMPER, crews will still be able to access a sufficiently large area of Mars and perform considerable research.

To highlight the advantage of the base-first strategy (as well as landing crews in capsules rather than in habitats), consider how much analogue research would cost if a new habitat had to be constructed and delivered to a different site for every single mission.

The DRA is designed as a precursor to settlement only in the sense that it would provide data and knowledge useful to future settlement plans; however, any material assets delivered to Mars would not be utilised for establishing PHP-M. In IMRS, however, the assets delivered to Mars during the missions form the seeds of a settlement. Components are intended and designed to last as long as possible, and be re-used across multiple missions.

This strategy enables infrastructure to be accumulated at the base, from which each successive mission will benefit. A communications, observation and navigation satellite will be positioned above the base in an areosynchronous orbit. Roads will be built. Transponders will be installed around the base to provide an accurate LPS (Local Positioning System) for use by vehicles and robots. Habitats, greenhouses, surface vehicles, power plants and other base components may be re-used, improved, developed, expanded, and integrated with each other and the surrounding landscape until the base can be permanently inhabited. Additional structures may be built from locally-sourced stone and metals, or by tunnelling underground or into hillsides.

A build-up of power-production resources in one location is of particular importance, and demonstrates the real advantage of this approach. Instead of sending shipments of power-production hardware to different locations on the planet, money can be saved by sending multiple shipments to the same location and thereby establishing a reliable and abundant power source at that location. The result is significantly improved energy security and reduced cost, which is crucial for long-term survival on Mars, and the same rationale applies to water, food, air, propellant, surface mobility, manufacturing, and so on.

By sending missions to distinct locations on Mars, it's true that more could be learned about Mars in the short-term. However, each mission would be almost equally risky and costly, and ongoing missions would be less likely. Sending hardware to a single location means that redundant backups of mission-critical components can be emplaced, thus reducing risk and cost with each successive mission. PHP-M can be established much quicker, and missions can be conducted from the settlement to other locations on Mars. The long-term result will be that more of Mars is explored and a new world will be opened up for human habitation.

The base-first approach obviously represents a huge cost saving; not only the cost of the hardware, but the considerably greater cost of sending it. Once the SHAB and CAMPER are at the IMRS it will only be necessary to deliver another MAV, crew and supplies in order to run each successive mission. If financial resources are available on subsequent missions, it will be possible to send backup components, or valuable additions to the base such as a greenhouse, additional power production or ISRU hardware, equipment for excavation or experiments, or other useful items of hardware that will facilitate PHP-M.

Focusing on a single location makes it crucial that an especially good one be selected, as will be discussed (see Site Selection).

4.4. Manufacturing and Materials

Amazing breakthroughs are currently occurring in materials and manufacturing that promise to be tremendously beneficial to space exploration.

For example, the combustion chambers of the SuperDraco thrusters in SpaceX's Dragon V2 capsule are the first ever 3D printed rocket engine components, printed using direct metal laser sintering of a material called

Inconel, which is an alloy of iron and nickel. The benefit of manufacturing parts this way is reduced mass, improved performance, and faster, cheaper production.

A related technological breakthrough is the development of *nanostructured materials*. Rather than being completely solid, these materials are comprised of tiny trusses at the nanoscale. The resulting materials can be stronger then steel, yet weigh significantly less; potentially even orders of magnitude less.

Some examples of truss forms printed in alumina are shown below:

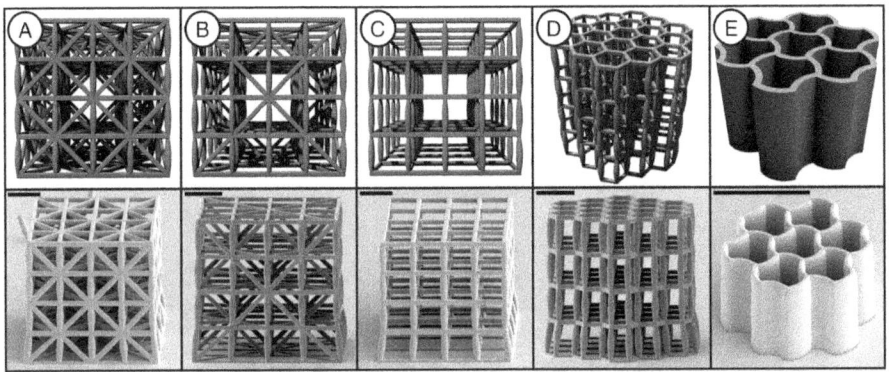

3D-printed microtrusses (Credit: Bauer et al.)

These micro trusses were 3D printed by German scientists using 3D laser lithography to create materials with strength comparable to steel, yet with one tenth the density (Bauer et al. 2014).

Nanostructured materials could become critical features of space missions, significantly reducing the mass of spacecraft components and thereby greatly increasing the utility of launch vehicles. In addition to mass reduction, this form of manufacturing can confer additional mechanical properties on components. Propellant tanks, for example, can not only be made much lighter, but also less permeable, and with enhanced resistance to microcracks.

One tenth the vehicle mass means one tenth the propellant requirement. Less propellant also means less tankage, further reducing the overall mass. If the vehicle can be made significantly lighter, then smaller or fewer rocket engines are necessary to propel it. Thus, even a modest reduction in spacecraft mass can ultimately translate to a significant reduction in IMLEO (Initial Mass in Low Earth Orbit), with a corresponding reduction in cost. Reduced vehicle mass can also translate to increased payload mass.

3D printing and nanostructured materials are receiving considerable attention and will improve dramatically in the coming years. Although the policy for development of the IMRS plan has generally been to base it, as much as possible, on present-day and near-term technologies with high TRL, these technologies are evolving so rapidly, and promise such enormous benefits, that a degree of reliance on their availability has been accepted.

Through 3D printing, robust and high-performing engine parts can be created at a fraction of the cost and time of traditional manufacturing methods.
- Elon Musk

5. Time

In this section, aspects of the plan related to time are reviewed. This includes the duration and timing of missions, an approximate schedule, and timekeeping during missions.

5.1. Duration

There are fundamentally two options for a HMM architecture:

1. Opposition-class or "short stay"

2. Conjunction-class or "long stay"

5.1.1. Opposition-Class Mission

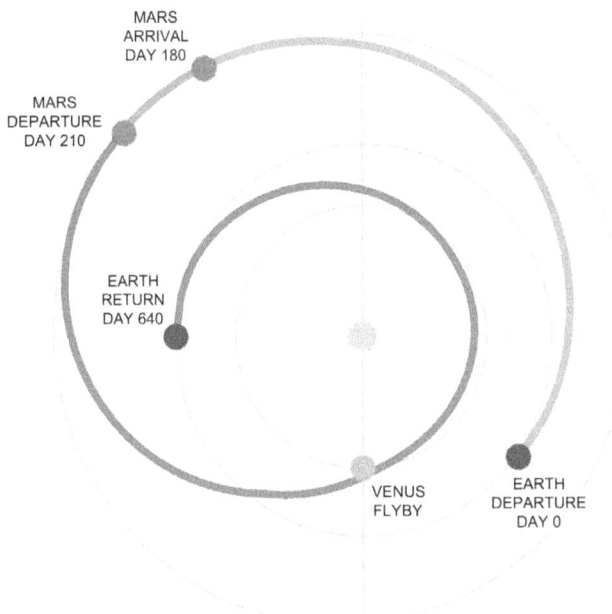

Opposition-class mission

An opposition-class or "short stay" mission involves a trip out to Mars, which takes about 6 months when Earth and Mars are optimally aligned. This is followed by a month or so on the surface doing science and exploration, then a launch from Mars and return flight home. However, by this time Mars and Earth will have moved to new positions in their orbits, and will no longer be optimally aligned. The return trip is therefore considerably longer; in fact, more than 1 year. The MTV swings by Venus on the return path, using its gravity to pull the vehicle into the correct trajectory to reach Earth.

Although actual numbers will vary depending on the timing of the launch, trajectory, and propulsion technology, the figures below are indicative of a schedule for a short stay mission with a 30-day surface element (Zubrin & Wagner, 1996).

Outbound flight	180 d
Surface mission	30 d
Return flight	430 d
Total	**640 d**
Total time in space	610 d
% time on Mars	5%
% time in space	95%

5.1.2. Conjunction-Class Mission

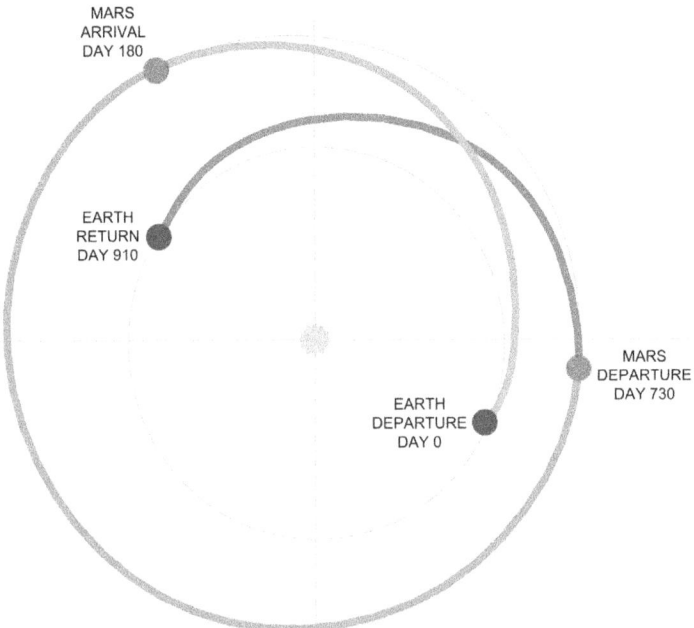

Conjunction-class mission

A conjunction-class or "long stay" mission similarly begins with a trip of approximately 6 months. The crew then remain on the surface while the two planets orbit the Sun along their respective paths until they're again optimally aligned approximately 1.5 years later. The crew then launch and return to Earth, again on a short-duration trajectory of approximately 6 months. The figures shown in the table below are indicative of a long stay mission schedule (Zubrin & Wagner, 1996).

Outbound flight	180 d
Surface mission	550 d
Return flight	180 d
Total	**910 d**
Total time in space	360 d
% time on Mars	60%
% time in space	40%

5.1.3. Comparison

A short stay mission may at first seem sensible and conservative for an initial mission.

One of the main benefits is that the quantity of supplies to be landed on Mars is greatly reduced, and ISRU would not be required for anything other than propellant. The amount of equipment to be delivered to Mars would also be reduced, as the crew would not have as much time available to do useful work on the surface. Since landing stuff on Mars is one of the most challenging aspects of the mission, this reduces complexity and cost; at least for the surface element. However, this is balanced by the fact that the space element would be more complex.

A short stay architecture minimises the length of time that the SHAB, surface vehicles, marssuits and other surface assets would be required, thus reducing the potential for any of these systems to fail during the surface mission. However, the amount of time that the MTV and its systems must be guaranteed for is then maximised.

Another argument that could potentially be made in favour of the opposition-class mission is that while we have decades of experience keeping astronauts alive in space, we have zero achieving this on Mars. However, the Martian environment is actually much healthier than interplanetary space, since Mars has only one quarter the radiation and provides a moderate level of gravity. The crew would receive a higher radiation dose on an opposition-class mission, and would be more severely affected by the deleterious effects of microgravity.

Thus, any supposed advantages of a short stay mission are entirely offset by the increased time spent in space, and at the cost of the actual purpose of the mission: to explore Mars.

The conjunction-class architecture is better for astronaut health, while also enabling them to achieve much more on Mars. Although the long stay option increases total mission duration by about 50%, the amount of time spent on Mars increases by a factor of approximately 18, accounting for 60% of the mission duration instead of just 5%. The amount of time spent in space is reduced by more than one third.

Most mission planners therefore agree that a long stay mission profile delivers a far greater ROI, which is why Mars Direct, Mars-Oz, the DRA and numerous other architectures are all based on conjunction-class missions. A 2.5-year mission represents significant risk and technical challenge; but then, so does a 1.75-year mission. If the short stay mission is achievable then so is the long stay one, but the pay-off is considerably greater.

Conjunction-class missions carry other important advantages over opposition-class missions:

1. The extended surface stay permits replanning of surface activities based on discoveries by the crew.

2. The crew can take a week or two to acclimatise to Mars's gravity and the SHAB's atmosphere, rather than having to begin work immediately.

3. The Δv is significantly less, reducing propellant requirements, MTV mass, and complexity of on-orbit construction.

4. Conjunction-class missions can be launched at each opportunity on a regular 26-month cycle. Opposition-class missions are much more sensitive to timing, and can only be launched when both Mars and Venus are in the correct positions in their orbits with respect to Earth.

5. Opposition-class missions take the crew close to Venus where solar radiation intensity is double that at Earth, and where warning time in the event of a CME (Coronal Mass Ejection) would be much less.

One advantage of the opposition-class mission, which should perhaps not be trivialised, is that it would afford the opportunity to observe Venus close-up and potentially deploy probes. However, it would be preferable in terms of safety and cost to design separate human missions to Venus optimised for that purpose.

5.2. Timing

Despite being referred to as a conjunction-class mission, the outbound and return journeys occur close to opposition of Mars, when Mars and the Sun are on opposite sides of Earth. This is approximately when Earth and Mars are closest, and occurs, on average, about every 780 days (or 26 months), a period called the *Earth-Mars synodic period*. Obviously it's quickest to travel between Earth and Mars when the distance is least.

Approximate timeline for a conjunction-class mission in relation to Mars oppositions

Due to Mars's elliptical orbit, every 15.8 years (on average) the distance between Mars and Earth at opposition reaches a minimum when it coincides with Mars perihelion, which is the point in Mars's orbit when it's closest to the Sun. This is called a *perihelic opposition*. The graph below shows the distance between Earth and Mars at opposition; the minima in 2003, 2018 and 2035 are the perihelic oppositions, and, as the graph shows, Mars is less than 60 Gm from Earth at those times (that's about 40% of the distance between Earth and the Sun).

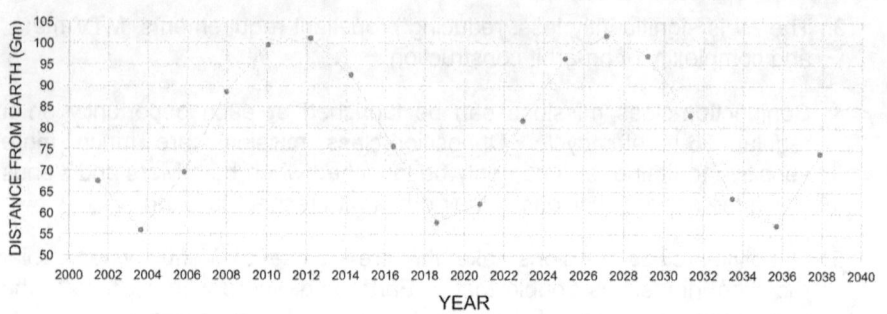

Distance between Earth and Mars at solar opposition (Credit: Jürgen Giesen)

Travelling to Mars at perihelic opposition minimises the trip time even more effectively because Mars is at its closest to Earth, not only within the 26-month synodic period, but within the 15.8-year cycle of oppositions.

The next perihelic opposition is in 2018, which is therefore a popular choice for a trip to Mars, especially since this will be close to the next solar minimum: the point in the 11-year solar cycle when sunspot and solar flare activity is at a minimum and therefore risk of a CME is correspondingly lower. This is why the Inspiration Mars flyby mission is planned for 2018.

To minimise the amount of time spent in space, and the amount of propellant required to get from Earth to Mars and back, the first crew (Alfa Crew) could be sent to Mars in 2018 and brought back in 2020. However, this is unrealistically soon, and the subsequent perihelic opposition is more practical. Sending them in 2033 and bringing them back in 2035 will achieve these things while also allowing sufficient time to build support for the mission, develop the international partnership, improve the TRL of the hardware, and run precursor missions.

There's no strict requirement to travel when Mars is close to perihelion; it would simply be advantageous for the first mission. Subsequent crews will spend longer in space (assuming equivalent propulsion systems) until the next perihelic opposition; however, those crews will benefit from the experience gained in previous missions, plus improvements made to the architecture and hardware. The first human mission is the most risky, during which astronauts and mission planners will have the least experience, and therefore it needs to be optimised for safety; hence, for that mission, the launch opportunity that will result in the shortest possible space element, and shortest overall mission duration, is preferred.

Therefore, the plan is that Alfa Crew will travel in 2033 (with predeployment in late 2030 - early 2031), Bravo Crew will travel in 2035, Charlie Crew in 2037, and so on.

5.3. Schedule

Knowing the mission duration and timing gives an approximate schedule for the first three IMRS missions:

Year	Program	Alfa Mission	Bravo Mission	Charlie Mission
2015-2020	Planning and design. Engage international and commercial partners. Develop collaboration frameworks.			
2020-2030	Precursor missions. Improve TRL of mission hardware. Select and train astronauts.			
2031		Predeploy supplies, MAV-1, SHAB-1, CAMPER-1, satellites.		
2032		Construct *Adeona*.		
2033		Crew outbound in *Adeona*.	Predeploy supplies, MAV-2, SHAB-2, CAMPER-2, greenhouse-1.	
2034			Construct MTV-2.	
		Crew return.		Predeploy supplies, MAV-

2035		*Adeona* parked on orbit.	Crew outbound in MTV-2.	3, greenhouse-2, additional base components.
2036				Refurbish *Adeona* and refill with propellant.
2037			Crew return. MTV-2 parked on orbit.	Crew outbound in *Adeona*.
2039				Crew return. *Adeona* parked on orbit.
2040-2100	Further missions and ongoing development and operation of the IMRS.			

5.4. Timekeeping

Timekeeping during IMRS missions will be based on Martian *sols* rather than Earthian days. A sol is the Martian diurnal (day-night) cycle. The length of a sol, about 24 hours and 40 minutes, is similar enough to a Earthian day that the crew, as well as mission controllers on Earth, can operate on a Martian diurnal cycle.

While on the surface of Mars the crew will live and work in harmony with the Martian diurnal cycle. This will be much easier and more energy-efficient than trying to implement an artificial diurnal cycle as will be necessary elsewhere in the Universe.

By also using sols instead of days while living in *Adeona* during the outbound trip to Mars, the crew can be spared any additional adaptation stresses on arrival in addition to the need to adapt to Mars's gravity. By the time they arrive at Mars, the crew will be used to the slightly longer day. In addition, by synchronising *Adeona*'s clocks with the diurnal cycle at IMRS, the crew will not experience "rocket lag" post-arrival.

For the return trip, the same rationale applies. By continuing to live in harmony with Mars's cycle instead of Earth's, the crew will be spared the stress of adaptation to a shorter diurnal period while simultaneously readapting to microgravity. Using the same onboard clock from the start of the mission to the end is also obviously simpler.

On arrival back at Earth the crew will need to readapt to the comparatively shorter Earthian day in addition to Earth's comparatively high gravity and thick, warm air. However, abundant medical personnel and equipment, as well as friends and family, will be available to help them readjust to Earth.

Living on Mars time has been done before. Mission planners working on the MER (Mars Exploration Rover) missions (*Spirit* and *Opportunity*) had special watches that operated on Mars time, and therefore, despite being still firmly on *terra firma*, began each day 40 minutes later than the one before.

Mission controllers will use clocks and watches synchronised with the diurnal cycle at IMRS. This is an additional advantage to focusing on a single location, as different clocks will not be needed for different locations on Mars. This will be important if they choose to live and work on Mars time like the MER mission controllers did.

A convenient unit for referring to time periods less than 1 sol is the *millisol* (abbreviated "msol"). Some convenient, approximate conversion factors:

$$1 \text{ sol} \quad = 1.0275 \text{ d}$$
$$= 24 \text{ h } 40 \text{ min}$$
$$40 \text{ msol} \quad = 1 \text{ h}$$
$$10 \text{ msol} \quad = 15 \text{ min}$$
$$1 \text{ msol} \quad = 90 \text{ s}$$

Converting the nominal mission schedule from days to sols:

Outbound flight	180 d	175 sol
Surface mission	550 d	535 sol
Return flight	180 d	175 sol
Total	**910 d**	**885 sol**
Total time in space	**360 d**	**350 sol**

6. Crews

This section discusses ideas pertaining to the design of crews for IMRS missions; in particular, the optimal crew size and roles necessary for successful HMMs.

6.1. Crew Size

The smaller the crew the better, as crew size affects mass of consumables, spacecraft, habitats, life support systems, surface vehicles and marssuits. Crew size therefore has a significant impact on mission cost.

The Mars Direct and MARS-Oz architectures require crews of only four; however, the Mars Semi-Direct, DRA and many other mission architectures specify crews of six.

Blue Dragon is also designed for six. There are several reasons for this:

- Because the DRA and many other architectures have been designed for a crew of six, existing research can be leveraged.

- Given the resources available, six is approximately the largest practical crew size while keeping the mission achievable and affordable.

- A crew of six permits a dedicated crew member for the most crucial functions, while also allowing for a degree of redundancy in skill coverage.

- B330 modules are designed to support six people with life support and ample volume. To use these modules and design the mission for less than six would be inefficient.

- A SpaceX Dragon V2 capsule is designed for seven people. It should therefore be large enough for six crew in MCP spacesuits, with some space allocated for samples during Mars ascent.

- Six crew members permits a degree of flexibility with team configurations.

Why not five, or three, or seven? The Apollo crews had three members. One advantage to having an odd number of people in the crew is that there's never a tied vote. However, it also means you can't use the buddy system. The "buddy system" is something children learn in school, but is also a good idea for HMMs. It simply means that everyone works in pairs. Your partner is your "buddy", and it's their job to keep you company, watch your back, and make sure you don't fall into a collapsed lava tube or step on a Martian rattlesnake. In turn, you do the same for them. This is important for safety as well as psychological health.

Naturally, the crew will not always be able to all work together. However, for safety and psychological reasons it is best to avoid leaving crew members to work alone.

Six people can be organised into three teams of two or two teams of three. This flexibility can be useful when organising shifts, EVAs and chores, it means no-one works alone, and safety and happiness is optimised.

6.2. Crew Roles

Each crew member on this mission will need to be capable of fulfilling multiple roles, from engineer to videographer to psychologist to gardener to commander. All skill sets are crucial to the mission and require redundant backups, and each astronaut on the mission must be trained in a range of functions.

It is proposed that the crew be comprised of two basic teams, namely Engineering and Science, each with three members. The engineers ostensibly have different roles; however, they must all also be able to fix any of the mission hardware, and therefore, to a large degree, be able to do each other's jobs. This applies equally well to the scientists.

Engineering Team	Flight and Mechanical	Handyperson, mechanic, pilot, surface vehicle operator.
	Mechatronics and Communications	Software, robotics, computers, electronics, antennas, multimedia.
	Chemical and Electrical	ISRU, ECLSS, propellant and power systems
Science Team	Planetary Scientist and Astronomer	Areology, planetology, navigation, site selection, cartography.
	Astrobiologist and Astrohorticulturist	Search for past/extant life, bio-experiments, food production, food systems.
	Medical and Safety	Health, fitness, nutrition, psychology, medical, safety protocols and drills.

Colour coding

As indicated above, each crew member will each be assigned a unique, distinct colour, which is theirs until the end of the mission. These colours are used for everything that belongs to that astronaut - their bunk, spacesuit, treat meals, clothes packs, towels, everything. Everyone in the crew will learn everyone's colours.

Apart from the advantages of not mixing up your towel with someone else's, one of the primary advantages of having distinct spacesuit colours is that it will be easy to identify who's who during EVA, particularly from a distance. Brown, orange and pink are excluded due to similarity with local colours on the Mars surface, as is black, which would be hard to see in the dark. Red should be ok, since, despite Mars being known as the "red planet", there's not a lot of actual red in the landscape.

6.2.1. Engineering Team

Flight and Mechanical Engineer

This person is not only an expert mechanical engineer, but also a mechanic. The role requires understanding, operating, maintaining and (if necessary) repairing, all space and surface vehicles, including *Adeona*, the MAV and the CAMPER.

Colour: White

Mechatronics and Communications Engineer

A master programmer, this role includes responsibility for all computing, communications and mechatronic equipment, including flight computers, on-board computers in the CAMPER, robotic systems, multimedia/web servers, personal computers and all communications hardware used in space and on Mars.

Colour: Blue

Chemical and Electrical Engineer

This person is responsible for operating, maintaining and repairing all chemical engineering hardware in the habitat and various spacecraft, including propellant systems, ISRU systems, ECLSS, waste disposal systems and plumbing. They're also in charge of monitoring and maintaining all power and electrical systems, both in the SHAB and in space and surface vehicles and machinery.

Colour: Purple

6.2.2. Science Team

Planetary Scientist and Astronomer

This role combines geology, planetary science, astronomy and cartography. On the surface of Mars they will study areology, areomorphology, areochemistry, areography, etc., and in space they will perform Earth, Mars and astronomical observation. They will be responsible for any telescopes used in space and on Mars, and producing detailed maps of the IMRS site and surrounding area.

Colour: | Red |

Astrobiologist and Astrohorticulturist

This role includes searching for and (if found) examining extant life on the surface of Mars. This person is also responsible for all food and food systems, and will experiment with food production in *Adeona*, and in the SHAB or (eventually) a greenhouse or laboratory on Mars.

Colour: | Green |

Medical and Safety Officer

This person is responsible for keeping the crew in good health for the duration of the mission. This critical role combines ship's doctor, personal trainer, psychologist and safety officer. It includes monitoring each crew member's physical and psychological health, including the effects of microgravity, radiation, and separation from Earth; ensuring crew members perform their exercises; providing nutritional advice; administering medications and treatments; monitoring solar flares; conducting JSAs (Job Safety Analysis); and developing and implementing safety protocols.

Colour: | Yellow |

6.2.3. Journalism Duties

If one more person could be squeezed into the mission it would be tempting to include a dedicated journalist and communicator, whose responsibility would be documentation and communications, including writing, photography, videography, blogging, vlogging, interviewing other crew members, and other forms of reporting.

This would be tremendously valuable to any space mission, as it would drastically increase engagement with the public (i.e. viewing audience) on Earth and thereby generating a greater ROI across the board, helping to justify the cost of the mission, generating a higher volume of feedback and good wishes for the crew, and thus improving morale and decreasing feelings of isolation. Most importantly, it would increase the connection people on Earth would have with the crew, the mission and Mars, which would inspire them and most likely spur further missions and settlement.

However, as ideal a dedicated journalist might be, it may be hard to justify for those first few missions. The alternative, which may in fact produce a better result, is to train all crew members in basic journalism and communications. Another approach would be to assign journalism duties to the Planetary Scientist and/or Astrobiologist during the space travel stages of the mission, since they may not be able to do as much science as they would like during that time. The two scientists could share the responsibility once on the surface.

There may simply be one or two crew members who are naturally more extroverted, expressive, and better suited to journalist-style duties, and who may voluntarily take on much of that role; a Carl Sagan, Brian Cox or Neil deGrasse Tyson, who could be both scientist and science communicator.

Another idea that could have an equivalent effect would be for all crew members to regularly report on their individual activities and experience via blogging and vlogging.

7. Site Selection

Because the intention is to send multiple missions to a single location in order to establish a permanent human presence on Mars, it's especially important that the location be a good one. The purpose of this section is not necessarily to identify a single optimal location for the IMRS, as that will require a more detailed analysis involving numerous experts from a variety of disciplines, but merely to highlight some of the salient characteristics such a location would have, and suggest a general approach.

There are several drivers in selecting an optimal location, which must be balanced:

- Availability of key location-dependent resources such as sunlight, heat and water. It may also be advantageous to be forward-thinking and also consider other location-dependent resources that will become more important in the future, such as wind and areothermal energy, minerals and metals, caves and lava tubes, and even tourist attractions.

- Terrain characteristics. For safety in landing, and ease and safety of surface mobility both in marssuits and surface vehicles, the location should be reasonably flat, level, and not overly dusty. In addition, it may be useful to have loose regolith available for piling around the SHAB for radiation protection and thermal insulation.

- Scientific interest. The best location will be close to sites that can help to answer scientific questions about Mars. Most importantly, has Mars ever hosted, or does it currently host, life; and, if so, what were, or are, its characteristics? Other questions relate to the presence of liquid water, nitrogen and other volatiles, and Mars's areological history. It's worth considering that, although planners will aim for the most scientifically important place, in reality anywhere on Mars will be of tremendous scientific interest to the first explorers.

Previous Mars missions, and the DRA, have primarily favoured scientifically important locations, in alignment with the intention to continue exploiting the scientific goldmine that is Mars, and the belief that human missions are likely to be few and therefore only a handful of the very best locations can be visited. However, the plan for the IMRS is that there be many human missions and expeditions as part of a broader program of settlement and research, and that ultimately all of Mars will be explored. With this in mind, selecting a site of maximum scientific value is less important than selecting one with maximum potential for supporting long-term habitation, because infinitely more science will be possible once a permanent presence is established. As has been pointed out before (Zubrin & Wagner, 1996), it's considerably easier and cheaper to provide long-range surface mobility on Mars than to run additional missions to new locations.

With settlement in mind, availability of critical resources such as solar energy and water are the strongest drivers for location selection, and to some degree these are at odds with each other. With increasing distance from the equator, the percentage of water ice in surface regolith increases, but, as on Earth, the availability of solar energy decreases. The best location choice will balance these, and is therefore likely to be somewhere between around 30-60° north or south. For a number of reasons related to terrain and Mars's unique orbital characteristics, as will be shown, north is preferable.

7.1. Sunshine

The section on Energy discusses the various reasons for favouring solar energy over other options. Due to the mission's reliance on it, intensity and availability of solar energy are especially important factors in location selection.

On Mars, as on Earth, the intensity of sunlight reduces with increased latitude. However, due to the eccentricity of Mars's orbit, northern latitudes enjoy milder weather than the southern. This can be advantageous for a northern situated solar-powered settlement, as the mass of the power system is reduced.

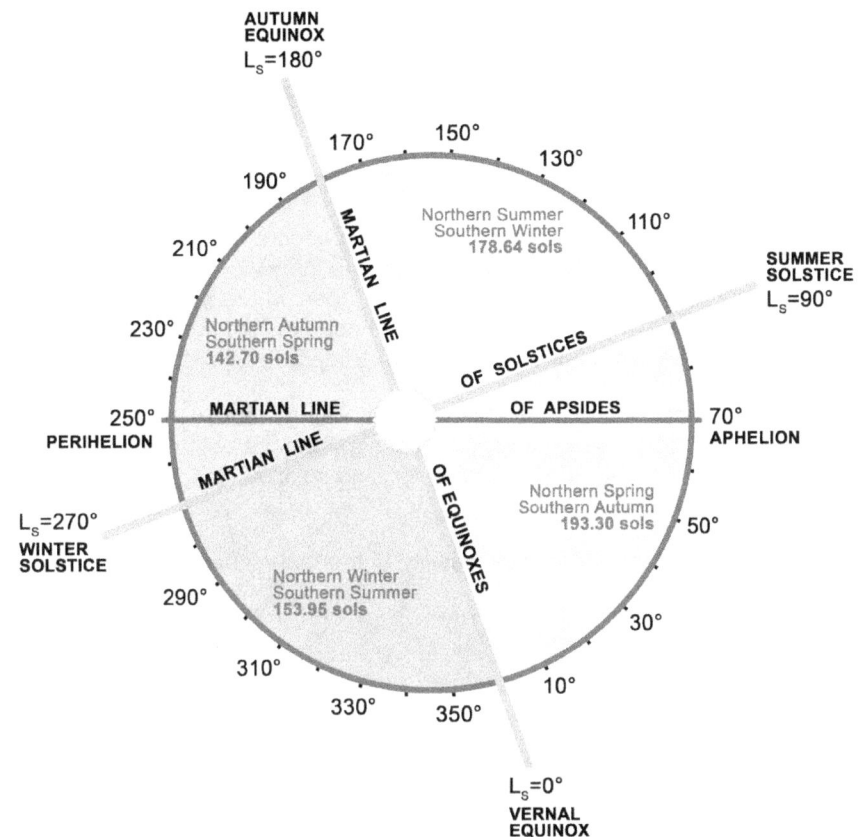

Seasons on Mars

Mars has an axial tilt of 25.2°, which is similar to Earth's axial tilt of 23.4°. This gives it a similar cycle of seasons. However, Mars's orbit is somewhat more eccentric than Earth's. Aphelion (when Mars is farthest from the Sun) occurs during the northern spring/southern autumn; and thus perihelion occurs during the northern autumn/southern spring. The effect is that seasons in the north are milder, whereas seasons in the south are more intense. The southern hemisphere has short, hot springs and summers, and long, cold autumns and winters. The north has long, cool springs and summers, and short, warm autumns and winters.

The net effect with regard to solar energy is that, in the northern hemisphere, there is less variation in the amount of solar energy received per sol during the course of the Martian year. The latitude 31°N has the highest minimum solar energy for a single sol during a Martian year (Cooper et al. 2010). In other words, the shortest sol of the Martian year at 31°N receives more sunlight than for any other latitude.

This affects energy needs as well as energy storage requirements. The longer the settlement must spend in darkness, the more solar energy it must be capable of storing. A settlement located in the south will require greater energy storage capacity to get through the long, dark winter, in addition to greater energy requirements for heating. The northern hemisphere requires less energy for heating during the milder winter, and the days are longer so more solar energy can be collected and stored.

For maximum exposure to sunlight, the location cannot be deep in a crater or chasm, as this would decrease the amount of time the PV (Photovoltaic) cells are sunlit each sol. Rather, it should be out in the open where it receives as much sunlight as possible. This is congruent with the need to land somewhere flat, and the open, flat terrain of the northern hemisphere supports this requirement.

Due to the low temperatures, one of the primary energy requirements of a habitat on Mars will be heating (see Energy for Habitats), and the best source of heat at the surface of Mars is the Sun. Selecting for high solar incidence simultaneously selects for warmth as well.

Two other factors that will affect availability of thermal management at the base are elevation and thermal inertia.

7.2. Terrain

7.2.1. Low Elevation

There are two main reasons for selecting a site at low altitude, and both are related to the fact that the atmosphere is thicker at lower elevations, as on Earth.

1. It will be warmer. The Martian atmosphere is warmest at lower elevations, as the thicker air functions as a thermal blanket.

2. The thicker atmosphere facilitates EDL technologies that use the atmosphere for deceleration; for example, parachutes, aeroshells and blunt-body capsules.

The altitude on Mars is generally lower in the north.

The flattest, smoothest, and also almost the lowest region of Mars is Vastitas Borealis, a vast area spanning the northern hemisphere that may have once been the floor of a huge ocean, usually referred to as "Oceanus Borealis". Vastitas Borealis is at a much lower altitude than the southern highlands. This is another important factor in favour of siting IMRS in the northern hemisphere.

The following map shows the topography of Mars based on data returned by the

MOLA (Mars Orbiting Laser Altimeter), showing Vastitas Borealis:

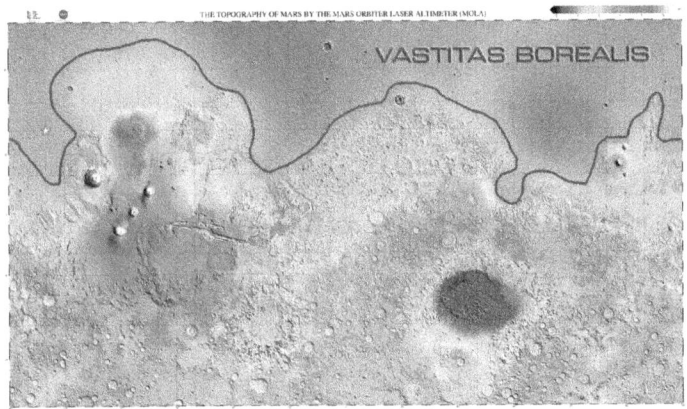

Mars topographical map generated from MOLA data. (Credit: NASA)

The lowest place on Mars is the huge impact basin in the southern hemisphere, called Hellas Basin, and for this reason some have proposed Hellas Basin as a suitable location for a base. It would be extremely interesting scientifically; however, it would not exploit other advantages of the northern hemisphere, the terrain is not very flat, and it's a dust trap. In fact, Hellas Basin is one of the main regions from where dust storms erupt.

Terraforming

One potential downside to choosing a location at low elevation for IMRS is that, if plans for terraforming proceed as envisaged, Vastitas Borealis may again become Oceanus Borealis, and the location of the IMRS — an extremely important historic site — will become submerged. Perhaps a watertight dome can be built over the site to protect it.

The various advantages of the north plus the benefit of living near water suggests that on a terraformed Mars, the bulk of the population may choose to settle along the shore of Oceanus Borealis.

7.2.2. Thermal Inertia

Thermal inertia refers to the ability of the terrain to retain heat. Small particles, such as dust, have low thermal inertia, i.e. they lose heat rapidly. Boulders and exposed bedrock have high thermal inertia, i.e. they retain heat longer. This is why people use exposed slabs of polished concrete indoors, as a thermal mass to reduce heating costs.

Thermal inertia at the location is important for several reasons:

- A very low thermal inertia implies thick dust, which would not retain heat and could impede mobility in marssuits and surface vehicles. Dust is also less useful for piling around and on the SHAB for thermal and radiation protection.

- A high thermal inertia is advantageous as the surrounding rock will serve to keep the base warmer after sunset, reducing energy requirements.

- A high thermal inertia may imply large boulders, which could impede mobility of surface vehicles and compromise safety when landing; or, very shallow bedrock, which could damage surface vehicles. Excavation, which may be desired for future development of the base, would also be more difficult.

- A medium thermal inertia implies loose regolith, which may be useful for piling around and on the SHAB, both for insulation to maintain habitat temperature after nightfall, and for radiation protection.

In other words, when it comes to thermal inertia, it may be a case of not too low and not too high. The following map shows thermal inertia generated from data obtained from TES (Thermal Emission Spectrometer), one of the instruments aboard the Mars Global Surveyor. Areas of intermediate thermal inertia, likely to be preferred, are shown in green:

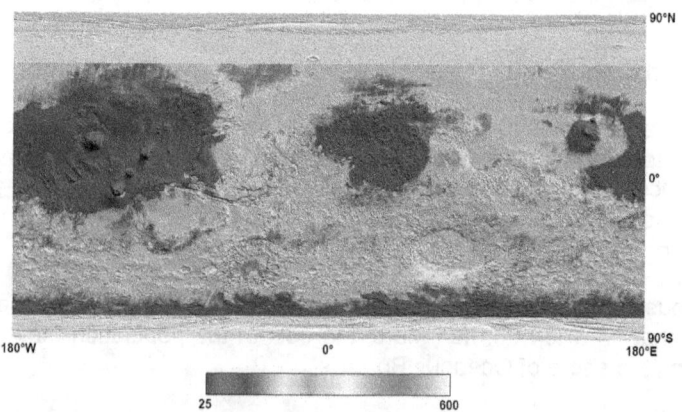

Mars thermal inertia map created from TES data (Credit: NASA; Arizona State University - Christensen 2001)

7.2.3. Surface Roughness

Choosing an especially smooth area is important for:

- Safe landings.

- Mobility in marssuits and surface vehicles.

Blue Dragon requires safely landing the MAV, SHAB, CAMPER and cargo capsules in approximately the same location. However, note that they will not be directly adjacent to each other, because landing any item will kick up large amounts of dust and debris; therefore, each item must be landed some distance, perhaps as far as a few hundred metres, from the others.

The CAMPER must be able to negotiate the terrain between these elements, and the surrounding area. In addition, the AWESOM (Autonomous Water Extraction from the Surface Of Mars) robot, discussed in In Situ Water Production, must be able to traverse the ground around the MAV.

In order to minimise the probability of landing on a large boulder, smoother terrain is preferred. Fortunately this corresponds with the usefulness of finding a location near a potential source of areothermal energy, because it's the locations that have most recently been covered in lava (and therefore have few craters) that are most likely to still be areologically active.

The downside of choosing an especially smooth area is that there may not be much of areological interest in the vicinity, which means more ground will have to be covered using the CAMPER rather than on foot. However, safety is a priority. The goal will be to locate the base within, at most, a few hours drive away from areologically interesting regions.

The following map shows surface roughness, with smooth areas appearing dark and rough areas appearing light. The smoothest region on Mars is to the north and west of Olympus Mons:

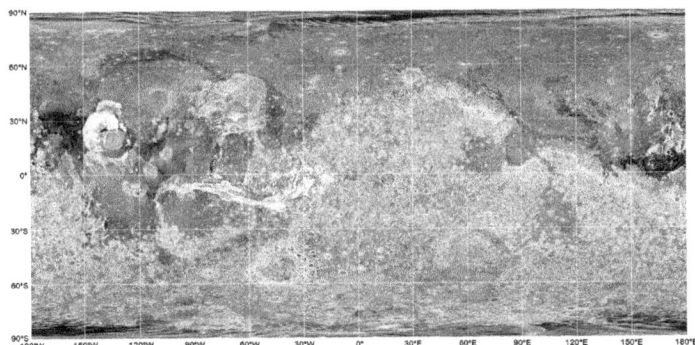

Mars surface roughness map (Credit: NASA; Arizona State University - Kreslavsky & Head 2002)

7.3. Water

One of the key goals of the IMRS is to make use of locally obtained water.

The small amount of water vapour in the Martian atmosphere may be sufficient to support recycling losses in the life support systems, but will not provide enough hydrogen for manufacture of ascent propellant. It will therefore have to come from the ground.

The following map has been generated using data from the GRS (Gamma Ray Spectrometer), an instrument aboard Mars Odyssey. Fortunately, it shows there's plenty of ground ice on Mars, comprising between about 1% and 70% of the regolith:

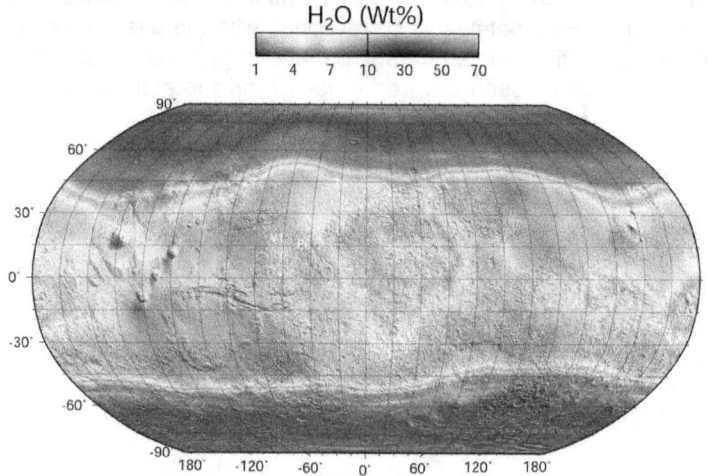

Mars water concentration map, generated from GRS data (Credit: NASA)

The ground on Mars is above 20-30% water beyond ~60° north and south, which is attractive. However, at these latitudes the solar incidence is somewhat low. If the goal is to remain close to 31°N, yet also have at least 10% water concentration in the regolith, these requirements may be balanced by selecting a location around 35-45°N. The areas with the highest water content near these latitudes are north-north-west of Olympus Mons, and north of Hecates Tholus (near Viking 2).

7.4. Candidates Sites

The northern hemisphere of Mars is likely to be strongly preferred for several reasons:

- More water, in the form of both ground ice and atmospheric water vapour.

- A higher minimum solar incidence.

- Milder climate.

- Lower elevation.

- Flatter and smoother terrain.

- Higher probability of areothermal energy sources and aquifers, which translates to a higher probability of finding extant life.

- More potential for discovery of mineral ores of the type formed by interaction between magma and liquid water.

The northern hemisphere also has disadvantages that are important to acknowledge:

- The smooth terrain, an advantage for safe landings and surface mobility, also means fewer sites of scientific interest; yet the base needs to potentially support decades of research. This could be addressed by improved surface mobility, such as longer range surface vehicles or aircraft.

- The lack of exposures through the crust formed by large valleys and craters could make it difficult to access potential mineral resources in the future.

Because the base is reliant on solar energy (see Solar), it should be located close to the optimal latitude of 31°N in order to minimise the mass of the power system. However, by going a little further north, more water can be accessed while still being in a fairly sunny latitude. One consideration is that, as latitude increases, the angle with respect to the MCOS (the areostationary satellite) becomes less favourable.

Somewhere in Vastitas Borealis may be optimal, as this region is low, flat, reasonably smooth, and has plenty of ice. However, it would be less than ideal to locate the base in the "middle of nowhere", as it were, because of the distance from interesting sites. If the various components can be landed with a reasonably high degree of accuracy, then the base could be positioned within driving range of features with greater scientific interest.

From the rudimentary analysis conducted in this section, several candidate locations are suggested:

Northern Amazonis Planitia, around 40°N 160°W

This may be the sweet spot in this region just northwest of Tharsis. There's abundant water at relatively low latitudes, with good solar incidence, and it's low and flat. This is one of the smoothest regions of Mars due to also being one of the most recently areologically active at only 100 million years old. This increases the likelihood that sources of areothermal energy will be found.

Arcadia Planitia is just to the north, which shows signs of near-surface ground water. From Wikipedia:

> In a lot of the low areas of Arcadia, one finds grooves and sub-parallel ridges. These indicate movement of near surface materials and are similar to features on Earth where near surface materials flow together very slowly as helped by the freezing and thawing of water located between ground layers. This supports the proposition of ground ice in the near surface of Mars in this area. This area represents an area of interest for scientists to investigate further.

Areothermal heat in combination with underground ice could produce aquifers at this location, which would be of tremendous interest as a resource as well as a potential home for extant Martian life.

This location is also close to Olympus Mons, one of the most famous and impressive features in the entire Solar System, which is both scientifically and aesthetically interesting. Olympus Mons will probably become the site of numerous settlements, purely for the views.

The thermal inertia is generally low to medium, indicative of dust, however, the data indicates there may be less dusty regions that could be suitable candidates for siting the base.

Utopia Planitia just north of Hecates Tholus, around 40°N 150°E

This area is flat and smooth, has about 7-10% H_2O, good solar incidence, low elevation, and moderate thermal inertia. The area is areologically very young and considered a candidate for areothermal energy. It's also within striking distance of Phlegra Montes, where radar probing has indicated large volumes of water ice below the surface.

This site is also close to older terrain in addition to inflow features of Cerberus Fossae, which would be of considerable areological interest.

Lyot, a crater in Vastitas Borealis at 50.8°N 330.7°W

Lyot is one of the more interesting features in Vastitas Borealis, 236 km in diameter, and the deepest point in the northern hemisphere.

Although not itself a smooth region, being in Vastitas Borealis means that IMRS could be located at a smooth place nearby, within easy driving distance of the crater. Alternately — again, assuming high-accuracy landings — IMRS could be located inside Lyot, providing a wealth of research for many decades.

Lyot is important primarily due to evidence of valleys, over 250 metres wide and tens of kilometres long, carved by flowing water in the relatively recent areological past. The following extract (Dickson, Fassett and Head 2009) emphasises the unique scientific value of Lyot:

> Regional drainage patterns suggesting liquid water stability at the surface are confined to early in the history of Mars, prior to a major climate transition to hyperarid cold conditions. Several later fluvial valley systems have been documented and are thought to have formed due to local conditions. Fluvial valley systems within Lyot crater have the youngest well-constrained age reported to date for systems of this size (tens of km). These valleys are linked to melting of near-surface ice-rich units, extend up to ~50 km in length, follow topographic gradients, and deposit fans. The interior of Lyot crater is an optimal micro-environment, since its low elevation leads to high surface pressure, and temperature conditions at its location in the northern mid-latitudes are sufficient for melting during periods of high-obliquity. This micro-environment in Lyot apparently allowed melting of surface ice and the formation of the youngest fluvial valley systems of this scale yet observed on Mars.

Thus, the higher air pressure in Lyot, an effect of the low elevation, can occasionally produce a localised climate at which water is stable in a liquid state at the surface. As liquid water is considerably more accessible than either ice or water vapour, this is an especially high-value resource for explorers and settlers. Even if liquid water is not or rarely available at the surface, it may be present close to the surface and accessible via drilling. Perhaps more importantly, liquid water may be indicative of life, as it always is on Earth. Lyot could potentially support decades of scientific research in areology, astrobiology and ISRU.

The following map shows known locations of hydrated minerals in Lyot, as detected by CRISM (Compact Reconnaissance Imaging Spectrometer for Mars) and OMEGA (Observatoire pour la Minéralogie, l'Eau, les Glaces et l'Activité, a.k.a. the Visible and Infrared Mineralogical Mapping Spectrometer). In addition, a close-up is shown of a region in south-east Lyot with a 10 km-long fluvial valley (i.e. ex-river bed) and a small, 1 km-wide crater. The surrounding area is quite smooth and more than large enough to accommodate the development of the IMRS.

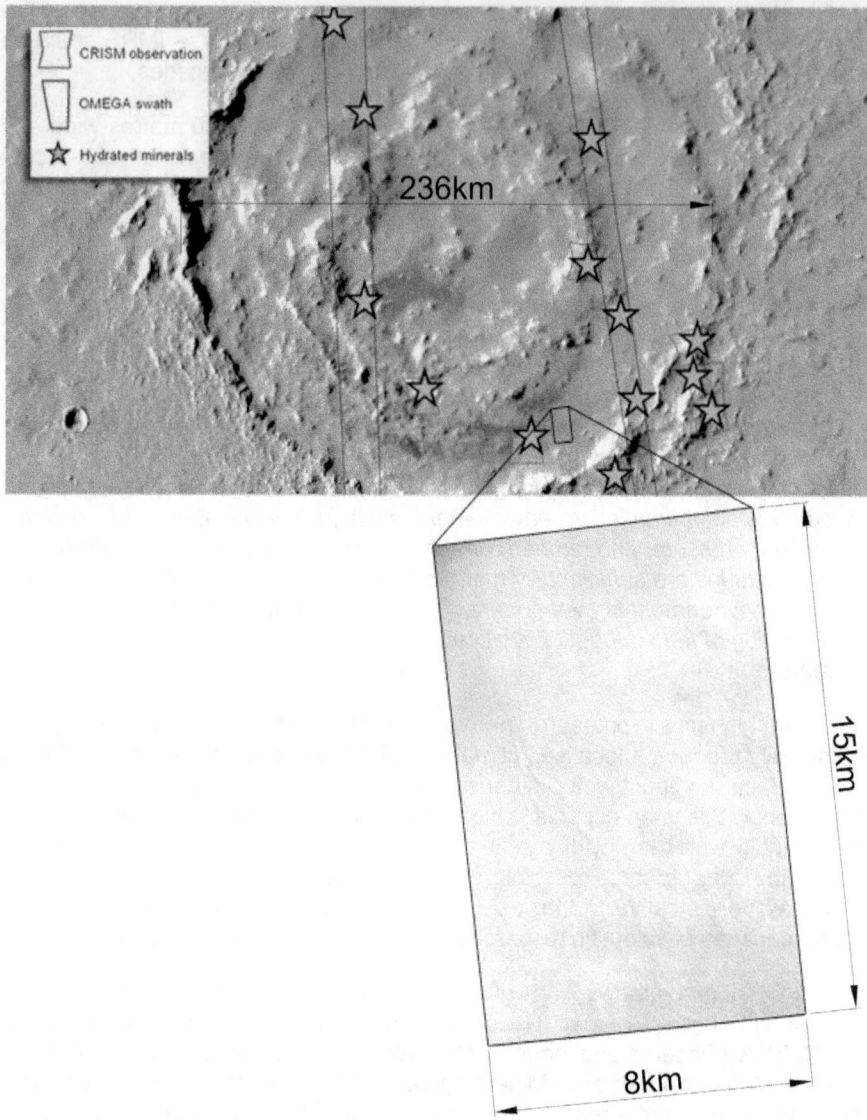

Map of Lyot showing hydrated minerals and close-up of fluvial valleys (Credit: JPL)

The landing ellipse for Curiosity was 20 km by 7 km. If this could be reduced to, say, 8 km by 5 km, it should be possible to land base components inside Lyot safely.

Lyot's latitude is somewhat high, reducing solar intensity, and locating a base inside a crater will also affect solar power production. However, if the site is considered sufficiently valuable, this problem could be solved with additional PV and energy storage hardware.

8. Resources

In this section, several ideas are presented related to the access, requirements, management and production of four resources fundamental to any human space mission: energy, air, water and food. Utilising local resources to manufacture methalox bipropellant is discussed separately (see In Situ Propellant Production).

8.1. Optimising Resource Usage

Space missions are effectively isolated from the environment of Earth where resources are comparatively much more accessible, and hence largely taken for granted. Resource management in space missions is paramount in order to ensure that sufficient quantities are available for the entire mission. In addition, the high cost of launching mass from Earth to space dictates that quantities of consumables to be brought from Earth are minimised.

There are three main strategies for resource management during the mission and to reduce the mass of consumables launched from Earth:

Rationing

Rationing rules must be followed by the crew in order to control consumption of life support resources within certain limits.

Rationing applies primarily to food and water, and simply means that a per-sol ration of each is allocated for each function (eating, drinking, washing, etc.). Crew members are required to strictly consume an amount equal to or less than their allocated ration, during each sol of the mission. In this way, a limited quantity of food and water can be made to last for the full duration of the mission. Naturally, a buffer is built into the supplies to allow some flexibility when required and for contingencies.

Other rules can be implemented to conserve energy consumption. For example, use of computing and communications equipment could be restricted to daylight hours when power can be drawn directly from solar panels instead of from less efficient sources such as batteries or fuel cells.

Recycling

This applies primarily to water and air, and is the most effective strategy for reducing consumable requirements.

Almost all water consumed by the crew can be recovered from blackwater, greywater, and the habitat atmosphere. On the ISS water is recycled with an efficiency of 90% or greater, significantly reducing the amount required for each mission.

The efficiency with which air can be recycled depends largely on how CO_2 is handled. In space it may be purged, or the O_2 component can be recovered using biological methods such as plants, or physicochemical methods such as LiOH (lithium hydroxide) canisters or SOECs (Solid Oxide Electrolysis Cell). On Mars, CO_2 from the SHAB atmosphere may also be purged, or it could be captured and recycled back into the ISAP (In Situ Air Production) system, especially if O_2 is being made from CO_2.

ISRU

ISRU refers to making use of resources from the Martian or space environment rather than transporting them from Earth. Examples of useful resources that could be obtained this way during early missions include water, hydrogen, oxygen, nitrogen, methane and methanol. Even the use of solar energy could be considered ISRU. As settlement progresses ISRU techniques will be used to produce bricks, glass, plastics, metals and everything else, as on Earth.

This approach is commonly referred to as "travel light; live off the land". Like the explorers who once trekked across the continents of Earth, those who were most successful typically only carried with them some of the resources they needed, obtaining the remainder from the natural environment.

The primary benefit of ISRU is a reduction in the amount of material that needs to be launched from Earth, which translates to reduced mission cost. The trade-off is the development of the equipment needed to obtain the resources, which must be carried instead.

8.2. In Situ Resource Utilisation

Blue Dragon proposes producing methalox bipropellant, potable water and breathable air from the local Martian environment.

Previous proposals for HMM architectures have been more conservative in terms of inclusion of ISRU, either excluding it altogether, or including only a moderate level. This seems rational, because provision and management of resources is critical for success of the mission, and ISRU has not yet been demonstrated "in the field", so to speak. However, because ISRU provides such a benefit in terms of reduction in launch mass and improved safety, it's worth mastering and incorporating as much as possible.

Although ISRU is yet to be demonstrated at Mars, this problem can be addressed by sending one or more robotic missions to do just that. A concept mission called

"Green Dragon", which is a Red Dragon capsule containing an integrated ISRU system that produces air and water as if for a habitat, would be a reasonably inexpensive method to achieve this. Another mission called "Gold Dragon" would test the operation of the MAV, including solar energy utilisation, water extraction and propellant production (see <u>Precursor Missions</u>). Although the cost of the Green and Gold Dragon missions is nontrivial, the benefits and ROI will be substantial, as demonstration of, and experience with, production of energy, water, air and propellant from locally obtained resources will be invaluable for future human missions and settlement.

ISRU is valuable not only for reducing launch mass, but also for improving safety. To highlight this, consider a situation in which the crew is reliant on water transported from Earth:

1. What if they run out of water? The amount required could be underestimated, perhaps because the conditions of living on Mars for an extended period necessitate higher requirements. What if there's a medical emergency? What if something causes water to be lost? (e.g. a micrometeorite puncturing a tank.)

2. What if the recycling equipment malfunctions and cannot be repaired?

3. What happens if the crew miss the return launch window for some reason and must stay on Mars for a further 26 months until they can attempt another launch?

If the crew have the necessary equipment to make potable water from the local environment, the crew will be in a much safer position; at least in terms of water.

Naturally, there are also risks associated with reliance on ISRU, such as, what if it doesn't work, or what if the equipment breaks down? The difference is that ISRU technology must be developed sooner or later anyway. Once developed, tested and refined, ISRU technology will ultimately produce a far more secure situation for the crew, and is essential for ongoing settlement.

ISRU of air, water, food and metalox will be discussed.

8.3. Energy

One of the primary challenges associated with any Mars mission is the design of power systems.

Energy is needed for the following purposes and more:

- ECLSS

- Lighting

- Heating

- Refrigeration and cooking

- Communications, computing and multimedia

- ISRU

- Airlock operation

- Recharging of marssuit and surface vehicle batteries

The primary energy sources under consideration for HMMs are nuclear fission and solar.

8.3.1. Nuclear

Nuclear fission energy has a clear benefit that accounts for its popularity in Mars mission designs: it can provide abundant, continuous power irrespective of weather or location. A fission reactor could provide ample power for ISRU processes, ECLSS and other electrical subsystems, and because nuclear reactors can provide power continuously, they do not require an associated energy storage solution. For this reason, many in the space community assume that HMMs should rely on energy from fission.

In Mars Direct, the ERV includes a nuclear reactor to provide energy to the ISPP system, which enables propellant production to occur rapidly. Once the ERV lands on Mars, a small robotic truck emerges with the nuclear reactor. The truck carries the reactor some way off, trailing electrical cable connecting it to the MAV, and places it in a nearby depression in order to prevent radiation products from irradiating the MAV, crew, or any other part of the base. This would ideally be a natural depression; or, one could be created using explosives. The reactor must be distanced from the MAV and surrounded by rock in this way because it will be unshielded.

Why is the reactor unshielded? On Earth, nuclear reactors are heavily shielded to absorb radiation, in addition to being encased in multiple containment shells in case of an accident. However, due to the high mass of these elements, it may not be practical to include them with reactors to be sent to Mars. To do so would make it even harder to land an ERV or MAV on Mars.

If the reactor is unshielded, even if moved some distance away from the MAV, it would create a no-go zone around it that would be unsafe for the crew to enter. The MAV may be protected from radiation, but the environment surrounding the reactor would not be. Irradiating the environment would interfere with scientific research by hampering exploration, polluting data, and possibly harming extant life.

Mars should be treated more responsibly than this. What if the reactor is parked near something of scientific interest? What if it needs attention for some reason? What is the long-term plan for decommissioning the reactor? A common argument against space exploration is: "We messed up one planet already, should we mess up another?" If these kinds of environmentally irresponsible practices are permitted, then such criticism is justified. Earth's environmental problems are the result of greed, short-term thinking, and wishfully assuming that the planet is big enough to absorb industrial pollution without considering if that policy will scale with population. Is the intention really to export the same dysfunctional thinking to Mars? We should have learned this lesson by now.

Our long-term goal must always be kept top of mind, which is to settle Mars and transform it into a beautiful new home for life. The primary benefit of the frontier is to invent better ways of living in harmony with each other and the environment, and the intention should be to bring the best of humanity to Mars and leave the worst behind. It's simply unacceptable to commence habitation of Mars with serious pollution. That kind of short-term thinking has well-proven negative effects.

This opinion does not mean fission should be completely ruled out as a power source. There are several new types of reactor currently in development that do not have many of the problems associated with the existing generation of reactors. Also, shielding mass could perhaps be significantly reduced by using new materials and manufacturing technologies, and, combined with the reduction in cost of access to space, future space nuclear reactors may be effectively shielded.

However, there is another significant problem associated with fission reactors on Mars: in the event of an accident, malfunction, or even routine maintenance, all of which are more likely in the unfamiliar and challenging environment of Mars, access to the reactor would be considerably more difficult, and there may not even be people around to do the work.

A nuclear accident on Mars combined with a global dust storm could effectively distribute radiation products over the whole planet. Regardless of how unlikely an event may be, if there is a better option for energy production then it really must be taken. Mars already has a difficult enough radiation environment without making it worse.

The IMRS plan includes nuclear energy sources in the form of ASRGs (Advanced Stirling Radioisotope Generator) powered by Pu-238; however, this energy is produced by radioactive decay, not fission. This type of nuclear power source is already present on Mars, as Curiosity is also plutonium-powered, as were Viking 1 and 2, and as the Mars 2020 rover will be. This form of nuclear power is significantly less dangerous than fission. In any case, strict safety protocols for managing radioactive materials of all kinds will need to be established for Mars.

8.3.2. Solar

Solar has been proven as a practical and reliable energy source on Mars. Along with many other Mars and space missions, the MERs Spirit and Opportunity were powered by solar energy. Spirit was active for 6 years, and Opportunity is still going after almost 10 years at the time of writing.

Research has shown that modern solar technology can match the mass and volume performance of nuclear energy on Mars (Cooper et al. 2010), and because of safety, environmental and policy issues, is likely to be preferred.

A rudimentary comparison of solar vs. nuclear fission:

Time to develop	Much less
Cost of development	Much less
Cost of manufacture	Much less
Cost of transportation	Less
Cost of operation	Less
Cost of decommissioning	Much less
Damage to environment (potential or actual)	Much less
Consistency of power output	Less
Dependence on location and weather	Much more

The main problem with solar panels is that they represent a fluctuating energy source that varies with diurnal and seasonal cycles, weather, and atmospheric dust levels. Dust can accumulate on panels, reducing their effectiveness, although with a human crew present this problem can be addressed with conventional push broom technology.

Solar energy at the surface of Mars is about half that of Earth. The solar intensity at the upper atmosphere is 43% compared with Earth, but this ratio is improved by the fact that Mars has a thin atmosphere and few clouds.

Solar energy is currently one of the largest areas of investment on Earth, and the technology is advancing rapidly. In 2011 investment in solar peaked at about $157 billion. Investment in rooftop solar peaked in 2012 at about $80 billion.

Solar is now a mainstream energy technology on Earth and will clearly become more so over the coming decades. Solar energy technology is suitable for Mars missions, with numerous advantages over nuclear.

Printable solar cells

One of the most important developments in recent years is the use of CIGS (Copper Indium Gallium Selenide) as a substrate for thin film solar cells. This material permits:

- printable solar material

- greater efficiency at converting solar energy to electrical energy

- greater precision and consistency with solar cell manufacture, which significantly reduces efficiency losses

- flexible solar panels that can be transported in rolls

The MAV and SHAB can carry rolls of flexible CIGS solar material that can be robotically unrolled onto the surface of Mars to collect solar energy in an efficient and simple manner, and with much less risk and difficulty than deploying a nuclear reactor.

8.3.3. Energy storage

For solar or any other intermittent energy source to be practical on Mars, energy storage solutions are necessary so that energy will still be available to the base during nights and dust storms. Some of the options suitable for Mars include:

- Batteries

- Graphene supercapacitors

- As chemical energy in the form of H_2, CO, CH_4, CH_3OH or other fuel for use in a fuel cell

Batteries may be preferable to fuel cells for several reasons:

1. Battery technology is evolving rapidly due to the proliferation of mobile devices and electric vehicles. The development of graphene Li-ion anodes has produced batteries that charge significantly faster and can store more energy. Batteries tend to be heavy, but the amount of energy stored per unit mass, and per unit volume, is increasing.

2. Batteries require minimal maintenance and are highly reliable. Fuel cells are comparatively more complex and can break more easily.

3. Fuel cells require a supply of O_2 for combustion, which, on Mars, costs energy to produce.

One benefit of a fuel cell is that it can potentially last much longer, whereas batteries can only be recharged a certain number of times.

Despite the benefits of batteries, inclusion of a CO fuel cell in the SHAB may be a worthwhile inclusion given that its ISAP system produces CO (see <u>In Situ Air Production</u>). This could be utilised as a nighttime energy source, assuming sufficient O_2 supplies.

Graphene supercapacitors are a technology currently in development, which will probably be available well before 2030, and may be far superior to batteries. They can store much more charge per unit volume or mass, recharge extremely quickly, and, unlike batteries, there's no limit on how many times they can discharge and be recharged. This technology is therefore receiving considerable interest due to its potential for use in electric cars, electronic consumer devices and other applications. Another significant advantage of graphene supercapacitors is that they're made of pure carbon, which means they could be manufactured from Martian air.

With these things in mind, the intention for the IMRS is to utilise graphene supercapacitors or graphene-anode batteries for energy storage, in addition to one or more CO fuel cells for nighttime power.

8.3.4. Future Options

Other energy resources available on Mars will become more important as settlement proceeds and technological infrastructure develops. The primary difference between the energy infrastructures on Mars and Earth will be that fossil fuels of any type, on which humanity has been so reliant, are probably unavailable on Mars. In addition, energy sources related to water, such as tidal, wave and hydroelectric energy, will be unavailable until terraforming is well advanced.

Wind energy

Considering the ubiquitous and constant wind on Mars, wind energy may become a major power source on Mars within a few decades. However, use of wind energy is unlikely in early Mars missions for several reasons:

- The air pressure on Mars is very low (less than 1% of Earth's), which means it doesn't generate sufficient force to produce much electricity. Comparatively, solar energy offers a better return in terms of energy produced per unit mass of hardware sent to Mars.

- Wind turbines operating in such a dusty environment may require high levels of maintenance that could be difficult or dangerous for settlers to provide.

- Early Mars missions will probably be located in low altitude areas, for warmth and water. The air may be thicker, but the wind will be slower.

Wind turbines on Mars may perform better at high altitudes.

Like solar and other renewables, wind energy technology is advancing at an exponential rate. New turbine designs, new materials and additive manufacturing should produce very lightweight, low-maintenance and low-friction turbines in the near future, and designs suitable for Mars are inevitable. The low gravity may permit the construction of very large turbines capable of powering settlements.

One advantage of wind energy is that turbines are simpler than solar cells and fabricated from less exotic materials; therefore, the capability to manufacture turbines from local Martian resources may be developed sooner than for solar cells. Due to the relative ease of manufacture, settlements could access wind energy more effectively than solar simply by producing and deploying turbines in quantity.

Areothermal energy

Areothermal energy may be available on Mars in regions of recent areological activity. If the IMRS is expanded into a larger-scale settlement, a nearby source of areothermal energy will be a huge advantage, potentially far more valuable than any other energy source currently under consideration.

Areothermal energy is both scientifically interesting and of immeasurable value to future settlements, capable of providing baseline power without the environmental issues associated with fission, or the need for energy storage subsystems as with solar and wind. Sources of areothermal energy may ultimately become the preferred locations for settlements on Mars.

Areothermal hotspots offer more than simply the potential for electricity generation. They could provide direct heating for settlements in addition to being associated with aquifers — valuable sources of liquid water.

Although postulated, the presence of areothermal energy is yet to be confirmed, and deep drilling equipment and/or sub-surface radar may be necessary to achieve this. In any case, it will be inaccessible to early explorers and settlers.

Planetary engineering expert Martyn Fogg has explored the potential for areothermal energy (Fogg 1996), and identified several regions that have been areologically active in fairly recently times:

- Cerberus Plains

- Hecates Tholus

- Medusae Fossae Formation

- Northwestern Tharsis

- Valles Marineris

Nuclear fusion

If developed into a mainstream energy technology, nuclear fusion may become important on Mars due to the relative abundance of deuterium (an isotope of hydrogen with both a neutron and a proton in the nucleus, also known as "heavy hydrogen"; ordinary hydrogen only has a proton).

In terms of atoms per litre of water, deuterium, a practical fusion fuel, is five times more abundant on Mars than on Earth, and 10,000 times more abundant than on the Moon (Zubrin 1996). It's important to note, however, that in terms of absolute quantities Earth has around 50 times as much deuterium as Mars simply because Earth has around 250 times as much water, plus deuterium on Earth is much easier to access since most of the water is in a liquid state. Therefore, deuterium will probably never be exported from Mars to Earth, although it may be an important resource to settlers on Mars and elsewhere.

One of the fusion reactions in which deuterium (^2D) is a reactant is:

$$^2D + {}^3He \rightarrow {}^4He + 1p + 18.3\ MeV$$

The second reactant in this equation, helium-3 (^3He), is far more abundant on the Moon than either Earth or Mars. If fusion ever becomes a major energy source, deuterium resources on Earth and Mars combined with lunar helium-3 could be valuable synergistic commodities in the triplanetary economy.

8.3.5. Energy for Habitats

Thermal control

One of the major energy requirements during the mission will be maintaining a comfortable temperature inside the habitats. For optimal human comfort and productivity the desired temperature inside the habitats will be in the range of about 295-298 K (22-25 °C, or 72-77 °F).

In the THAB, as in the ISS, the primary requirement will be cooling. From the Wikipedia entry for ISS:

> *The large amount of electrical power consumed by the station's systems and experiments is turned almost entirely into heat. The heat which can be dissipated through the walls of the station's modules is insufficient to keep the internal ambient temperature within comfortable, workable limits. Ammonia is continuously pumped through pipework throughout the station to collect heat, and then into external radiators exposed to the cold of space, and back into the station.*

Appropriate solutions for thermal management may already be built into the

B330. If not, solutions could be adapted from those developed for the ISS. The thermal environment of interplanetary space is similar to that in LEO, although it will become cooler farther from the Sun.

Although the priority in the THAB will be cooling, due to the low temperatures on the Martian surface the primary energy requirement for the SHAB may be heating.

The surface temperature on Mars varies between 130 K and 308 K, with an average of about 218 K, which is about the same as the interior of Antarctica. The SHAB will therefore need to maintain an internal temperature approximately 80 K warmer than the external environment, on average. This is a significant temperature gradient, and one that must be maintained constantly.

As in the THAB, the operation of electrical and electronic equipment in the SHAB will produce plenty of heat, which will help to reduce the amount of energy required for heating. Also, the shell of the B330 provides significant thermal insulation. This could be augmented with regolith piled around the SHAB, which would also provide additional protection from radiation.

When external temperatures are higher, the interior of the SHAB may, like the THAB, require cooling. The thermal control subsystem must therefore be capable of controlling temperature in both directions. Good design can produce both passive cooling and heating effects.

Further analysis is required to determine the actual energy requirements for thermal control in the habitats.

SHAB energy subsystem

Although an uninterrupted power supply is not required for the MAV, it is a requirement for the SHAB because the ECLSS must operate continuously. As discussed, the SHAB will be powered by solar energy, which means an energy storage solution is necessary.

Some percentage of the solar energy collected each sol will be stored for use during the night. However, due to efficiency losses in energy storage, as well as limits on how much energy can be stored with the available equipment, it's important to minimise the amount of energy to be provided from storage.

Energy required by the SHAB during the Martian night, when no solar power is available and the SHAB is reliant on stored energy, can be reduced significantly if a simple rule is implemented specifying that the crew sleeps or at least rests while it's dark, i.e. wake at dawn each sol and do all work in daylight hours. With this regulation in place, power for lighting, computing, communications, laboratory experiments, etc. will mainly be required during the daytime when solar power is directly available.

The SHAB subsystem requiring the most power during the night will be the

ECLSS, partly because other electrical loads will be minimised, but mainly because energy requirements for heating will be higher than during the day. Communications will also require constant power. Even when the crew is not using the communications channels, telemetry data will be streaming back to Earth continuously.

The B330 has two arrays of solar panels, which will be deployed on arrival of the SHAB at Mars in order to power the ISAP, cameras, antennas, environmental sensors and other systems. For additional power, as mentioned, rolls of PV material could be unrolled on the surface.

8.4. Air

Breathable air is required in the THAB, SHAB, CAMPER, capsules and marssuits. The crew will always be wearing their marssuits inside the Dragon capsules, although they should only need to breathe from the suits during ascent from Mars.

1. **THAB:** The B330 will be inflated and pressurised with breathable air very similar to the Earthian atmosphere. Its built-in ECLSS will maintain the air quality during the outbound and return missions, in much the same way as the ISS: scrubbing excess H_2O, CO_2, CH_4, NH_3 and other contaminants from the habitat atmosphere, and topping up with O_2 and N_2 as required to maintain the correct partial pressures.

2. **SHAB:** As discussed below, the SHAB atmosphere will have a pressure of around 53 kPa. Compared with Earth, this is slightly greater than half the atmospheric pressure at sea level, or the same as at about 5.5 km altitude. The SHAB will include a compact and lightweight ISAP system to manufacture breathable air from the local Martian environment, which is necessary for maintaining the desired operating pressure within the SHAB, compensating for losses due to airlock cycling and other causes, and as a source of O_2 for marssuits.

3. **CAMPER:** The CAMPER's atmosphere will match the SHAB's. It will include a compact, lightweight ECLSS similar to that found in a Dragon V2, and may be able to top up or freshen its air supply by connecting with the SHAB.

4. **Marssuits:** Marssuits will provide astronauts with pure O_2 to breathe, and will scrub exhaled CO_2 with something like LiOH canisters. The suits' O_2 tanks will be refillable from the ISAP system, with refill ports available on both the interior and exterior of the SHAB.

5. **Einstein:** This capsule will initially contain normal Earthian air. This will mix with Adeona's atmosphere, which has virtually the same composition, after docking. When the crew use this capsule for Mars descent, it will

initially contain air from *Adeona*. When they arrive at Mars surface and pop the hatch, that air will escape and the capsule will fill with Martian atmosphere. However, the crew will be breathing pure O_2 from their suits, having already prebreathed aboard the MTV prior to descent.

6. **Kepler**: When the crew enters *Kepler* it will contain Martian atmosphere, but the crew will again be wearing their marssuits and breathing pure O_2. Upon docking with *Adeona*, its atmosphere will mix with air from inside the THAB before the crew transfer into *Adeona*.

7. **Newton**: As this capsule is launched from Earth, it will initially contain normal Earthian air. When it docks with *Adeona*, this air will mix with the spacecraft atmosphere, which has virtually the same composition. The crew will then enter the capsule and descend to Earth surface.

Maintenance of atmosphere pressures

The atmospheric pressure in the THAB, SHAB and CAMPER will reduce over time due to:

- Purging of waste gases (e.g. CO_2).

- Leaks. Although designed for zero leaks, in practice some small leaks are possible.

- Experiments.

- Airlock cycling (for the SHAB and CAMPER).

- Refilling of marssuit and CAMPER tanks from the SHAB's supply.

Pressure loss will be managed in the same way as on the ISS. Oxygen and buffer gas are kept in separate tanks so that both the partial pressure of each can be precisely controlled. If the partial pressure of either falls below a given threshold, then the atmosphere is replenished from the respective tank.

As explained below, the buffer gas in the SHAB will not be pure N_2 but a mixture of N_2 and Ar in approximately equal proportions.

Ventilation and circulation

Because warm air does not rise in microgravity, ventilation and circulation will be important in the THAB, as in the ISS. Air around laptop computers, for example, can heat up considerably. Consistent air flow through the ISS is maintained via the use of fans.

Within the SHAB, even though it will be operating in a gravity environment, ventilation will still be important. Any volume can produce hot and cold spots if air circulation is inadequate. Effective circulation is also important to ensure that accumulated CO_2, NH_3 and other contaminants in the habitat atmosphere are effectively scrubbed by the ECLSS.

8.4.1. THAB Atmosphere

The atmosphere inside the THAB will be very similar to normal Earthian conditions, with one atmosphere (101.3 kPa) pressure comprised of 21% O_2, about 78% N_2, plus small amounts of CO_2 and H_2O. This matches the ISS, Shuttle, Vostok, Mir and other space stations and spacecraft.

There are many reasons why an Earth-like atmosphere is preferred for use in space:

- Discounting hypogravity effects (such as hot air not rising), the behaviour of the atmosphere is predictable in terms of its interactions with organisms and equipment.

- It reduces the cost of equipment inside the spacecraft. Rather than needing to design equipment for a specialised atmosphere, as was necessary with Apollo, Skylab and many other early spacecraft, with an Earth-normal atmosphere many ordinary COTS items developed for Earth can be used in the spacecraft without alteration.

- No adaptation is required for astronauts when moving between the spacecraft and Earth. Physiological effects of long-term exposure to a reduced pressure or low-oxygen atmosphere are not a consideration.

- No time or energy is required to pressurise or depressurise capsules for crew transport to or from Earth.

- A high percentage of buffer gas greatly reduces fire risk.

The B330 is designed for an Earth-normal atmosphere, since one of its proposed applications is space tourism.

8.4.2. SHAB Atmosphere

The advantages of a normal Earthian atmosphere apply equally well to the SHAB as they do to the THAB. However, a good case can be made in favour of a reduced atmospheric pressure for the SHAB.

If a custom SHAB was being built, one reason would be the potential for reduction in its mass. The higher the internal atmospheric pressure of the SHAB, the stronger and heavier it needs to be in order to contain that pressure. However, the mass of the SHAB should be minimised as much as possible due to the difficulty of landing such a heavy object on Mars. In this plan it doesn't matter, because by using a COTS item instead of a custom-built habitat, the SHAB's mass will be the same regardless of internal air pressure.

A far more compelling reason for a reduced atmospheric pressure in the SHAB is the potential for a ZPB (Zero PreBreathe) protocol for EVAs.

Zero Prebreathe Protocol

Spacesuits typically use pure O_2 as breathing gas, which simplifies the PLSS (Personal Life Support System) and thereby reduces the suit mass. Minimising suit mass is crucial if astronauts are to healthfully conduct many extended EVAs, as they will wish to do so while on Mars; hence early marssuits will probably also only provide O_2.

Spacesuit pressures have historically ranged from 26 kPa to 57 kPa. Spacesuits providing higher pressure are sometimes referred to as "Zero Prebreathe Spacesuits" (ZPS), as they permit an astronaut to transition to the suit without needing to spend any time prebreathing. Prebreathing involves breathing pure O_2 for a period of time, from 30 minutes to several hours, at atmospheric or mid-level pressures in order to purge N_2 from the blood and other tissues. This is necessary with low-pressure suits to prevent DCS (decompression sickness, also known as "the bends").

ZPSes for Mars are highly desirable, as they will result in considerable saved time and much greater convenience for the crew, who will wish to go outside as often as possible. It's far easier to achieve a ZPB protocol with MCP suits, as these can support a higher pressure while maintaining good mobility. Gas-pressurised suits, on the other hand, require very low air pressure in order to facilitate effective mobility.

The amount of prebreathing required is based on the decompression ratio (also known as the "R value"), which is the ratio between the partial pressure of buffer gases in the spacecraft or habitat atmosphere, and the total pressure in the spacesuit:

$R = P_{BG} / P_S$

where:

P_{BG} is the buffer gas partial pressure in the spacecraft

P_S is the total suit pressure

The IMRS crews will be spending approximately 535 sols on the surface of Mars, with a plan to go on EVA up to four sols per week each, which is around 300 each (technically up to 600 if the EVA is divided into two shifts). Due to this large number, a ZPB protocol is only practical for an R value of 1.0, because risk of DCS is cumulative. From research by Lockheed (Campbell 1991) (highlight added):

> *An R value of more than 1.0 may have a risk of decompression sickness.*
> *This risk increases with increased R value. The risk of decompression*
> *sickness during a mission is statistically cumulative over a number of*
> *decompressions during that mission; therefore, the mission duration and*
> *frequency of EVA must be considered when determining an appropriate*
> *value for R for a given mission. A statistical analysis of cumulative risk*
> *shows that for R=1.22, the risk of decompression sickness after 10*
> *decompressions is 7%, while for R=1.4 the risk is 37%. Therefore,*
> *cumulative risk is an important criterion for long-term planet surface missions*
> *and it increases rapidly as R is increased above 1.0. If the decompressions*
> *are not separated by enough time to eliminate previously formed gas bubble*
> *nuclei from body tissues, the risk of decompression sickness on subsequent*
> *decompressions is also greatly increased.*

Fortunately, it is possible to design an atmosphere for the SHAB that will permit an R value of 1.0 in order to support a ZPB protocol without risk of DCS, which will benefit the astronauts' health while saving considerable time. This is a very strong reason in favour of a non-standard atmosphere in the SHAB.

Atmospheres in early spacecraft had low total pressure, low oxygen pressure, or both. However, research conducted since then has more clearly defined the limits for artificial atmospheres.

Oxygen pressure

The minimum partial pressure of oxygen required to support human physiology is 16 kPa. Some researchers recommend a minimum partial pressure of oxygen of 18 kPa for long-duration space missions (Larson & Pranke 1999); however, 16 kPa is known to be safe, and, as will be shown, permits a ZPB protocol without risk of DCS, which is ultimately better for the crew.

The partial pressure of oxygen in airline cabins ranges from 16 kPa to 18 kPa, hence many people are familiar with this oxygen level. 16 kPa O_2 is equivalent to an altitude of about 2400 m. Most people can ascend to this altitude without difficulty, although altitude sickness may occur above this level if the person is not acclimatised. More than 140 million people live permanently at an altitude of 2400 m or greater. Astronauts can be conditioned for an O_2 pressure of 16 kPa by training in a hypobaric chamber, or at this altitude.

Total pressure

A pure O_2 atmosphere introduces an unacceptably high risk of fire, such as that which occurred in the Apollo 1 Command Module. The upper limit of oxygen concentration with regard to fire safety is not clearly defined, but 30% is considered reasonable. If the O_2 partial pressure is 16 kPa, the minimum total atmospheric pressure must therefore be 53.33 kPa, or slightly more than half the atmospheric pressure at sea level on Earth.

Carbon dioxide pressure

NASA specifies a maximum CO_2 concentration in spacecraft atmospheres of 0.7% (Toxicology Group 1999). A CO_2 concentration of 1% may cause drowsiness, with more serious symptoms occurring at higher concentrations. A typical concentration in normal spacecraft operations is 0.5%, which is a reasonable design goal for the SHAB. The target CO_2 partial pressure is therefore approximately 0.27 kPa.

Water vapour pressure

NASA specifies a RH (Relative Humidity) of 30-70% for spacecraft atmospheres; i.e. an average of about 50%. At 295-298 K and a pressure of 53.33 kPa, the saturated water vapour pressure is 2.62 kPa. The target average water vapour partial pressure will be 50% of this, or 1.31 kPa.

Buffer gas pressure

Ignoring trace quantities of other gases present in the SHAB atmosphere, the required buffer gas pressure can be calculated by subtracting the partial pressures of O_2, CO_2 and H_2O from the total pressure.

$$P_{BG} = P_{TOTAL} - P_{O2} - P_{CO2} - P_{H2O}$$

$$= 53.33 \text{ kPa} - 16.00 \text{ kPa} - 0.27 \text{ kPa} - 1.31 \text{ kPa}$$

$$= 35.75 \text{ kPa}$$

The buffer gas in the SHAB will not be pure N_2, as in the THAB and ISS. The ISAP system, as described in the next section, extracts N_2 and Ar together from the Martian atmosphere for use as buffer gas. Because the Martian atmosphere is 1.9% N_2 and 2.1% Ar, this gives:

$$P_{N2} = (1.9 / (1.9 + 2.1)) * 35.75 \text{ kPa}$$

$$= 16.98 \text{ kPa}$$

$$P_{Ar} = (2.1 / (1.9 + 2.1)) * 35.75 \text{ kPa}$$

$$= 18.77 \text{ kPa}$$

With this buffer gas pressure, an R value of 1.0 is possible if the marssuits can provide an equal pressure. This is indeed possible, as discussed further in the section on <u>Marssuits</u>.

Proposed SHAB Atmosphere

The proposed atmosphere is therefore as follows:

Constituent	Partial pressure (kPa)	Fraction (%)
Argon (Ar)	18.77	35.2%
Nitrogen (N_2)	16.98	31.8%
Oxygen (O_2)	16.00	30.0%
Water vapour (H_2O)	1.31	2.5%
Carbon dioxide (CO_2)	0.27	0.5%
Total	53.33	100.0%

Auditory effects

One effect of the different gas mix in the SHAB will be a slightly slower speed of sound compared with normal Earthian air. Argon and O_2 molecules are heavier than those of N_2, which tends to reduce the speed of sound, but the higher heat capacity ratio of Ar partly counteracts this. The net effect is a slight decrease.

At 298 K and 50% RH:

Speed of sound in THAB/Earthian air = 347.25 m/s

Speed of sound in SHAB/CAMPER air = 333.45 m/s

A lower speed of sound will produce proportionally lower frequencies for fixed wavelengths such as those produced by vocal chords. Hence, the astronauts' voices will sound slightly deeper than usual, with a frequency reduction of about 4%.

A more obvious effect of the reduced air pressure will be a noticeable reduction in loudness. While this will make the SHAB more peaceful with regard to the whirring of fans and pumps, the astronauts' voices will also sound much quieter and they may need to develop a practice of speaking louder than usual to be heard (a habit they'll unlearn once back in the THAB). This has safety implications, but since the astronauts will never be more than 9 metres away from each other when in the SHAB, the risk of not being heard is low.

Physiological effects of reduced O_2 and reduced gravity

Mission doctors must consider the effects on the heart and cardiovascular system of a reduced O_2 partial pressure in combination with reduced gravity.

Acclimatisation to reduced O_2 pressure at altitude is characterised by an increase in pulse and breathing rate as the heart works harder to deliver available oxygen to the cells. Yet, in contrast, the physiological effects of reduced gravity include a *decrease* in heart rate and arterial pressure, as the heart doesn't need to work as hard to pump blood against the force of gravity. The low Martian gravity may therefore serve to counteract the effects of low O_2 partial pressure in the SHAB. More research is necessary to determine the combined effects and potential health risks.

Modifications to the B330

The ECLSS, and perhaps other components of the B330, may require modifications necessary to support the lower air pressure, a buffer gas containing argon, and continuous operation for 1.5 years on Mars. In addition, it should include backups for all subsystems, enabling any to be taken offline for repairs or maintenance as required.

8.4.3. CAMPER Atmosphere

Aspects of atmosphere design relevant for the SHAB apply equally to the CAMPER. The goal is to provide a safe and healthy atmosphere and support a ZPB protocol, while keeping the mass of the CAMPER as low as practical.

There are several advantages if the CAMPER atmosphere has the same composition and pressure as the SHAB:

- The CAMPER can be made lighter than it would need to be with an internal pressure of one atmosphere.

- No adaptation or prebreathing will be required when transitioning between the SHAB and the CAMPER.

- The same ZPB protocol for EVA can be used for both the SHAB and the CAMPER.

- The same equipment (such as laptops, scientific instruments and sensors) can be used in both environments.

With the SHAB and CAMPER having the same atmospheric composition it may also be possible to connect the CAMPER directly to the SHAB; for example, with a person-sized hatch to make it easy to transfer between them without needing to suit up or go through the airlocks; or perhaps simply with a pipe to refresh and replenish the CAMPER's atmosphere.

8.4.4. In Situ Air Production

One of the goals for the SHAB is that it include an ISAP system that will manufacture breathable air for the crew and replace losses. The inclusion of such a system will obviate the need to bring breathing gas from Earth, and relax constraints on air supply.

The primary constituents of the Martian atmosphere are CO_2, N_2, Ar, whereas those of Earth's are N_2, O_2 and Ar. Therefore, all the elements required to make breathable, Earth-like air are readily available from Martian air.

After landing, the SHAB's systems will be remotely activated from Earth, initiating the ISAP unit and filling the SHAB's tanks with O_2 and buffer gas. As the SHAB is being sent one launch window earlier than the crew, there will be about 20 months to achieve this, in addition to testing the SHAB's other systems, before the crew leave Earth.

The SHAB will most likely be inflated and fitted out on Earth prior to sending it to Mars, in order to permit installation of all equipment, fixtures and fitting in advance, thus sparing the crew this task. If the SHAB does not arrive at Mars fully inflated, however, the ISAP unit could be used to inflate it to the required operating pressure before the crew arrive.

The DRA proposes partial ISAP in which N_2 is extracted from the Martian atmosphere for use as a buffer gas. However, it is energetically cheaper to simply use the gas that remains after dust, CO_2, H_2O and toxins are removed from Martian air, as this is mostly N_2 and Ar, and both are perfectly good buffer gases. Argon comprises 1% of air on Earth and is non-toxic and unreactive, and there's no practical reason to separate it from the N_2, which would require additional energy due to the low boiling points of N_2 and Ar (assuming fractional distillation).

The ISAP Process

Step 1: Mars atmosphere is drawn into the system through a dust filter.

Step 2: Water is removed from the gas mix via adsorption by zeolite 3A (see In Situ Water Production). The captured water is stored in the SHAB's water tank.

Step 3: As in the ISPP system, CO_2 (about 96%) is separated from the remainder of the gas mix via condensation.

Step 4: The CO_2 is electrolysed into O_2 and CO using SOECs. Both products are stored.

The mixture of gases that remains after CO_2 removal is mostly comprised of N_2 and Ar, with small amounts of other gases, including O_2 and water vapour, plus a few undesirables: O_3, CO, and NO. (NO is technically non-toxic, but rapidly oxidises to toxic NO_2 when mixed with O_2.) Trace amounts of noble gases such as neon, xenon and krypton, which are inert, non-toxic and harmless, may also be present.

Step 5: The mixture of gases is first passed through an ozone scrubber (an inexpensive COTS item), which reduces the O_3 to O_2.

Step 6: The remaining gases are then passed through an ordinary automobile 3-way catalytic converter (also an inexpensive COTS item), which converts CO into CO_2, and NO into N_2.

Step 7: The result is a safe buffer gas comprised predominantly of N_2 and Ar, with small amounts of O_2 and CO_2, and traces of inert noble gases. This mixture can be combined with additional O_2 to provide breathing gas for the SHAB as per the design atmosphere described above. The small amount of CO_2 in the buffer gas will not be an issue, as any excess will be readily scrubbed out by the ECLSS.

SHAB ISAP Schematic

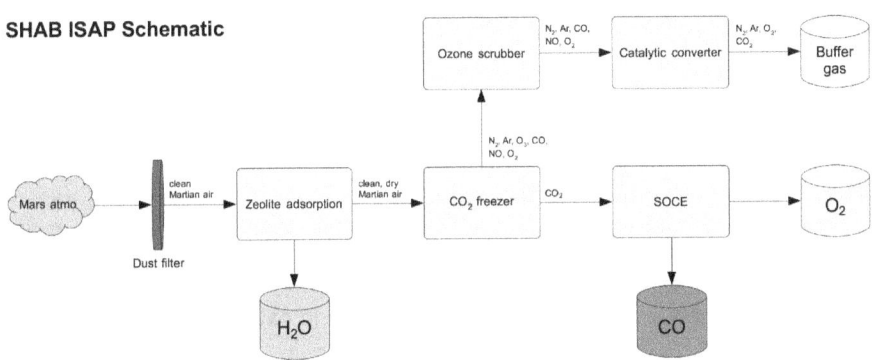

In Situ Air Production schematic

More research is required to complete a design for this system. It is necessary to determine rates of breathing gas consumption and production (including recycling efficiency); gas separation techniques; the unit's mass, volume and power requirements; and how it will integrate with the ECLSS.

Because the atmosphere is 96% CO_2, it's possible that much more O_2 than buffer gas can be produced, depending on the speed and efficiency of the SOECs. If 1 tonne of atmosphere was fully processed with 100% efficiency, the result would be:

- About 33 kg buffer gas

- About 352 kg O_2

- About 615 kg CO

The SHAB atmosphere is about 30% O_2, 70% buffer gas by weight. (This is approximately the same proportion as by volume, because a 50-50 mix of N_2 and Ar has about the same overall molecular weight as O_2.) Therefore, for each 33 kg of buffer gas, only about 14 kg of O_2 is required for habitat air, leaving a potentially large surplus.

There are several uses for surplus O_2:

- Marssuit breathing gas.

- In combination with CO, this may be an effective means of energy storage. O_2 and CO may be produced during the day using solar energy, then recombined in a fuel cell during the night to produce power for the ECLSS and other systems.

- Inflating weather balloons. As O_2 is lighter than CO_2, balloons filled with O_2 will rise in the Martian atmosphere.

- If H_2 is brought from Earth as described in the DRA, it could be combined with O_2 in order to make water.

8.5. Water

Water is one of the most important resources required for a HMM, fundamental to life support in addition to being a source of hydrogen for ascent propellant. Calculating exact water requirements is difficult, as water is present in food, air, and all human wastes, and can be used for purposes as diverse as drinking, food preparation, hygiene, spacesuit cooling, and food and oxygen production.

Approximate calculations of water requirements are attempted, using the following assumptions:

1. Both the THAB and SHAB include a Water Recovery System (WRS) similar to the WRS in the Tranquility module in the ISS. The WRS on the ISS can recover 93% of the water it receives; however, NASA has expressed an intention to improve the efficiency of water recovery to greater than 95%. The WRSes on the THAB and SHAB are assumed to have an efficiency of about 95%, which is probably realistic given the development time frame.

2. All water used will ultimately become available to the WRS:

    - Water will be separated from human metabolic wastes, including perspiration, urine and faeces, via centrifuge and lyophilisation (freeze-drying). Residual solids will be stored in waste containers as powder. This process recovers the water while also making the waste biologically and chemically stable and minimising waste storage mass and volume.

- Any water used for spacesuit or marssuit cooling will be retained in the suit rather than vented to the environment, and returned to the WRS for recycling.

3. Water is not used as a source of O_2; rather, O_2 is fully recovered from exhaled CO_2 in the THAB, or produced from Martian atmospheric CO_2 in the SHAB.

4. The astronauts won't lose any mass during the mission. In actuality, they probably will, including some water.

Ascent propellant

The MAV uses Martian H_2O as a source of hydrogen for methalox ascent propellant. This is discussed separately (see In Situ Propellant Production).

8.5.1. Space Element

Life support

The following values (ed. Hanford 2004) are used for calculating approximate water requirements for life support. Values have been converted to kilograms per person per sol (kg/p-sol), to account for the fact that the crew will be living on Martian time during the mission.

Potable water consumption (kg/p-sol)	
IVA (Intra-Vehicular Activity)	4.016
EVA, additional	5.918
Human metabolic water production (kg/p-sol)	
IVA	0.354
EVA, additional	0.395

The THAB will include 1-2 EVA spacesuits for emergency use in the event that external spacecraft inspections or repairs are necessary, or for fun. EVAs during the space element will probably be infrequent and brief, and hence additional potable water consumption for EVA doesn't need to be considered. However, the crew will be exercising up to about 80 millisols (about 2 hours) per sol to counteract the effects of microgravity, which, for sake of a first-order analysis, can be counted as effectively equivalent to EVA in terms of physical exertion.

Hence, potable water consumption per sol is approximately:

4.016 kg/p-sol + (0.08 * 5.918 kg/p-sol) = 4.489 kg/p-sol

About 2 kg/p-sol will come from crew drinks, with the balance coming from the water content of food, plus additional water used in meal preparation.

For a crew of six:

4.489 kg/p-sol * 6 p = **26.93 kg/sol**

It's worth noting that the values shown for additional water consumption and production as a result of EVA were originally calculated for gas pressurised spacesuits, which require much greater physical exertion than MCP suits. They are therefore likely to be high; however, they will suffice for this exercise.

Metabolic water production

A small amount of water is produced by the body as a byproduct of energy production, and this additional water is released into the spacecraft environment as part of the crew's metabolic waste. This water is captured by the ECLSS and purified, and therefore represents a small gain that counteracts water consumption.

Metabolic water production will be:

0.354 kg/p-sol + (0.08 * 0.395 kg/p-sol) = 0.386 kg/p-sol

For a crew of six:

0.386 kg/p-sol * 6 p = **2.32 kg/sol**

Hygiene water

Different values are recommended (ed. Hanford 2004) for hygiene water requirements in different space habitation situations.

Hygiene water needs (kg/p-sol)

	ISS/Transit vehicle	Early planetary base	Mature planetary base
Urinal flush	0.31	0.51	0.51
Oral hygiene	0.38	0.38	0.38
Hand wash	-	4.19	4.19
Shower	-	2.79	2.79
Laundry	-	-	12.81
Dish wash	-	-	5.59
Totals	**0.69**	**7.87**	**26.27**

If the goal was to minimise water usage as much as possible then the preferred values for the space element would be those for a transit vehicle. However, the values shown were developed for an ISS-style habitat, which is much more compact than the B330. In fact, they are so conservative that metabolic water production would almost exactly balance recycling losses, causing the net water loss during the mission to be virtually zero. While that would be good, the THAB would probably not smell too nice after a few months and the astronauts would feel quite grubby. These things would cause adverse psychological effects.

The B330 has the capacity to carry a reasonable supply if water, and, in fact, the intention is to line the interior of the THAB with water tiles in order to provide additional radiation protection. While this could potentially be achieved using some other material, having a moderately abundant supply of water provides the important psychological benefit of being able to wash hands and bathe (it's hard to shower in microgravity, but at least the astronauts can bathe with a wet cloth). In addition, this provides a safety margin that can be accessed simply by skipping hand washing and bathing.

Therefore, for these calculations the values for an early planetary base are used for both the surface and space elements of the mission.

For a crew of six:

7.87 kg/p-sol * 6 p = **47.22 kg/sol**

Spacesuit cooling

As discussed in the next section (<u>Surface Element</u>), water may be required for cooling spacesuits and marssuits, depending on their design. However, as mentioned, EVAs probably won't be a major feature of the space element.

Spacesuits would only be required infrequently and briefly, and therefore consume minimal water for cooling.

Other

Additional water may be required for:

- Research, such as growing plants in space. This amount is difficult to estimate without a clear definition of the experiments.

- Medical incidents. Requirements will depend on the incident, but an average value per incident may be on the order of 1-5 kg. The frequency of incidents is difficult to estimate, although during the space element they should be very low.

A nominal water budget of **0.25 kg/sol** is initially allocated for these uses, to be modified later as details of the mission evolve.

Recycling

The total water consumed during the space element, per sol, is approximately:

26.93 kg/sol + 47.22 kg/sol + 0.25 kg/sol = 74.40 kg/sol

The quantity of grey water received by *Adeona*'s WRS includes metabolic water:

74.40 kg/sol + 2.32 kg/sol = 76.72 kg/sol

The quantity of purified water produced from this will be approximately:

95% * 76.72 kg/sol = 72.88 kg/sol

The net rate of water loss is therefore:

74.40 kg/sol - 72.88 kg/sol = **1.52 kg/sol**

Contingency in case of abort to orbit

The DRA includes contingency water in *Adeona*'s supplies in case all or part of the surface mission is aborted and the crew need to remain in, or return to, the orbiting MTV until the TEI window opens.

This contingency quantity is equal to approximately 1.5 years' supply, i.e. the full duration of the surface mission. However, the rate of water consumption will not be identical to that required for the surface mission because of the lower requirements for spacesuit cooling and and other miscellaneous uses such as medical and ISFP.

Surplus contingency food and water, in addition to waste, will be jettisoned before TEI in order to reduce mass.

Contingency in case of equipment malfunction

Adeona's equipment, including the WRS, will naturally be designed for the lowest possible failure rate. However, a failure rate of zero is impossible to guarantee, particularly during a space mission. Equipment failure rates typically increase over time, and this equipment will have a minimum lifetime of 2.5 years; potentially much longer if reused for multiple missions.

If the WRS fails, grey water can be stored until repairs are performed, after which the system can be operated with a higher-than-usual rate to recycle that water. However, it may still be prudent to have a contingency buffer just in case.

At this early stage it's difficult to determine exactly how large such a contingency buffer needs to be, but 1 week's supply will serve as an estimate for now. This would cover a situation in which the water recycling unit was offline for up to 1 week.

In such a situation, bathing and washing of hands would be suspended in order to conserve water usage. This would reduce daily hygiene water usage to transit vehicle levels:

0.69 kg/p-sol * 6 p = 4.14 kg/sol

This buffer would therefore be:

(26.93 kg/sol + 4.14 kg/sol + 0.25 kg/sol) * 7 sol = **219 kg**

Total

The total duration of the mission is about 885 sols. Because of the possibility that the crew may need to spend this entire period living in *Adeona*, the total water requirement depends on this duration rather than simply the total duration of the space element (350 sols).

The minimum amount of water that *Adeona* needs to carry is therefore:

(1.52 kg/sol * 885 sol) + 219 kg = **1564 kg**

The estimate may be somewhat low, especially if water is required as a source of O_2. However, it will suffice for now.

As all surplus food, water and waste will be jettisoned prior to TEI, it's also necessary to calculate how much water must be retained for the return trip. This is:

(1.52 kg/sol * 175 sol) + 219 kg = **485 kg**

8.5.2. Surface Element

EVA Schedule

The amount of time spent on EVA pertains directly to water requirements, as potable water consumption, metabolic water production, and requirements for marssuit cooling water all increase linearly with EVA time.

One approach regarding to duration and frequency of EVA is a typical work-week: 5 days per week, 8 hours per day (ed. Hanford 2004). This makes sense for short missions, when time spent on EVA must be maximised in order to reap the greatest scientific ROI for the mission. However, for a long surface mission the pressure to maximise EVA time will be much less, and having all astronauts outside at the same time would compromise safety. Based on this, the following system is proposed for the IMRS:

1. For 6 sols per Martian week there are two EVA shifts, nominally of 160 msol each (~4 h). The maximum total EVA duration per person per sol will therefore be 320 msol.

2. The maximum number of crew members on EVA at any one time will be four. For safety, there must always be at least two crew members in the SHAB, communicating with crew members on EVA and with MCC, writing and recording reports, cleaning and maintaining the SHAB, and so on.

3. One sol per week (Martian equivalent of Sunday) the whole crew remains together in the SHAB for "family day" group activities (e.g. Sunday lunch, movie), rest and relaxation, and personal free time.

Therefore, in each week, each crew member nominally has up to 4 sols on EVA, 2 sols working in the SHAB, and 1 sol for free time. The average amount of time spent on EVA per crew member will therefore be:

0.32 sol/sol * 4 sol / 7 sol = **0.183 sol/sol**

Life support

The same base values as for the space element are used for calculating water requirements for the surface element. However, a larger amount of time will be spent on EVA.

Potable water consumption per sol:

4.016 kg/p-sol + (0.183 * 5.918 kg/p-sol) = 5.099 kg/p-sol

For a crew of six:

5.099 kg/p-sol * 6 p = **30.59 kg/sol**

Metabolic water production

Metabolic water production will be:

0.354 kg/p-sol + (0.183 * 0.395 kg/p-sol) = 0.426 kg/p-sol

For a crew of six:

0.426 kg/p-sol * 6 p = **2.56 kg/sol**

Hygiene

The rate of hygiene water usage will be the same in the SHAB as in the THAB, as the values for an early planetary base are used. Therefore, the requirement will be **47.22 kg/sol**.

Marssuit cooling

There are a variety of methods for cooling conventional space suits (ed. Hanford 2004):

- Circulating water (requires about 0.57 kg/h of water)

- Radiator (requires about 0.19 kg/h of water)

- Cryogenic spacesuit backpack (requires no water)

As yet, it is unknown which of these the IMRS marssuits will employ, or if some new or alternative method will be used. For the purpose of these calculations a radiator is assumed for marssuit cooling, and thus an estimated water requirement of 0.19 kg/h of EVA while on Mars.

In Martian time this is equal to:

0.19 kg/h * 24 h/d * 1.0275 d/sol = 4.685 kg/sol

The estimated rate of water use for marssuit cooling is therefore:

4.685 kg/sol * 0.183 sol/p-sol * 6 p = **5.14 kg/sol**

Other

As for the space element, additional water is required for research and for medical incidents. In both cases, more water will be required during the surface element. Experiments will be conducted in ISFP (e.g. growing vegetables), which will require water, regardless of the growing method. Areological and astrobiological laboratory research may also require water. Medical incidents will be more frequent considering the large amount of time spent working outside.

Therefore, an additional water budget of **1 kg/sol** is allocated for these requirements. Again, this value may be adjusted as details about surface activities become available.

Recycling

The total water consumed during the surface element, per sol, is approximately:

30.59 kg/sol + 47.22 kg/sol + 5.14 kg/sol + 1 kg/sol = 83.95 kg/sol

The quantity of grey water received by the SHAB's WRS includes metabolic water:

83.95 kg/sol + 2.56 kg/sol = 86.51 kg/sol

The quantity of purified water produced from this will be approximately:

86.51 kg/sol * 95% = 82.18 kg/sol

The net rate of water loss will therefore be:

83.95 kg/sol - 82.18 kg/sol = **1.77 kg/sol**

Contingency in case of equipment malfunction

As for the THAB, it will be prudent to have in reserve 1 week's supply of water to allow time to repair the WRS in the event of a malfunction. Hygiene water usage would be reduced to transit vehicle level during such a situation, and EVA would be suspended. Therefore the buffer needs to be:

(30.59 kg/sol + 4.14 kg/sol + 1 kg/sol) * 7 sol = **250 kg**

Total

The duration of the surface mission is approximately 535 sols. Adding this 1 week contingency buffer, the estimated minimum quantity of H_2O required for the surface mission is:

(1.77 kg/sol * 535 sol) + 250 kg = **1197 kg**

Ideally, this is the minimum amount that the SHAB's ISWP system will produce and store in the tanks before the crew depart Earth. The rate of water production necessary to accumulate this amount of water in ~584 sols is:

1197 kg / 584 sol = **2.05 kg/sol**

8.5.3. In Situ Water Production

The DRA proposes transporting hydrogen to Mars for the purposes of making water for life support. O_2 obtained from atmospheric CO_2 would be combined with the H_2 to produce H_2O and energy. However, there are a variety of nontrivial problems associated with transporting and storing H_2, which are discussed in greater detail in <u>Propulsion</u>. In addition, storing H_2 in the SHAB could incur a risk of explosion.

The plan for Blue Dragon is to obtain all the water needed for the surface mission from the local Martian environment. Although this is technically challenging, it is achievable and worthwhile. Transporting hydrogen across 100 Gm of interplanetary space to a world where it's already present in abundance would be highly inefficient, both in terms of the cost of transportation, and the investment in engineering solutions with zero long-term value. The capability to extract water from the Martian environment is a fundamental requirement for long-term habitation of Mars, and therefore must be developed eventually anyway.

Water on Mars is almost entirely present in the form of ice, with approximately 5 million cubic kilometres having been identified at or near the surface. A small amount of water vapour is also present in the atmosphere. Various hydrated minerals and clays have also been identified, and, although the quantity of water in this form is currently unknown, it may be significant.

Ideally, one or more precursor missions will be implemented to test and develop ISWP equipment before running human missions (see <u>Precursor Missions</u>).

Obtaining water from the atmosphere

The concentration of water vapour in the Martian atmosphere is very low, however, it has the advantage of being much more accessible.

At the University of Washington in 1998, researchers designed a device called a WAVAR (WAter Vapour Adsorption Reactor), which extracts water from the Martian atmosphere via adsorption into zeolite 3A (Grover et al. 1998; Adan-Plaza et al. 1998). The 885 kg WAVAR unit was designed to produce 3.3 kg of water per sol, a quantity determined to be sufficient for crew water requirements during the surface element of the NASA DRA.

As calculated above, the minimum water requirements for Blue Dragon may be less then this amount. A unit similar to the WAVAR may therefore be ideal for an IMRS mission.

Assuming that this rate of water production is achievable, a WAVAR unit of this size attached to the SHAB could produce 1927 kg of water before the crew leave Earth. This is greater than the minimum required for the mission, which would give a high confidence in water security (again highlighting the benefit of predeploying the SHAB). The WAVAR would continue to produce another 578 kg during the crew's outbound trip of 175 sols, and another 1766 kg of water during the surface mission, for a total of 4271 kg of water.

This is more than three times the calculated minimum requirement, and would go a long way towards total water security. If the WRS suffers a serious failure, this quantity of water could last up to 17 weeks. Therefore the WAVAR unit (as described in 1998) will be adequate for providing water for the surface element, and presumably an even better unit can be designed and built using 21st century engineering.

The gap between departure of Alfa Crew and the arrival of Bravo Crew is only about 8 months, allowing less time to fill the tanks, but there will still be plenty.

The value of water at the IMRS cannot be overstated, and as much as possible should be produced and stored. Each mission's cargo should include water tanks for building up a stockpile at the site, and improving water security for each subsequent mission. An abundance of water will enable increasingly ambitious experiments in food production, and will therefore be even more valuable once greenhouses are added to the base assets from the second or third mission onwards. Beyond the minimum requirements outlined in the previous section, surplus water can serve a range of functions to improve comfort and security at the base:

- More elaborate cooking and meals

- More extensive ISFP

- Laundry

- Showers

- Cleaning the SHAB

- A valuable backup source of O_2 and H_2

- Feedstock for methane production

Obtaining water from the ground

For the SHAB, sufficient water for the first few human missions can be obtained from the atmosphere. However, far more can be obtained from the ground, and in order to support food production, longer missions and larger crews, and to provide optimal water security, this capability should be developed as early as possible.

For the MAV, the quantity of water necessary for propellant production is too large to rely on atmospheric sources. Blue Dragon therefore introduces a system called AWESOM, which is a mobile robot that traverses the nearby terrain, extracting water from the top layer of regolith, and delivering it to the MAV.

Research has shown that a substantial fraction of the ice in the top layer of regolith may be liberated using microwave radiation (Ethridge & Kaukler 2012). The frequency of the microwaves can be tuned to preferentially heat water molecules rather than dirt.

As discussed in Site Selection, the ground at the proposed location for the IMRS should be at least 7% water by mass, and ideally about 10%. It should also be reasonably flat and free of rocks to support mobility of the AWESOM robot. The robot will traverse the ground around the MAV collecting water, and returning to the MAV to unload at the end of the sol, or whenever its tank is full.

The Curiosity rover is powered by a plutonium-fuelled RTG (Radioisotope Thermoelectric Generator) known as the MMRTG (Multi-Mission RTG), which produces about 110 W of electricity from about 2000 W of heat. While an RTG would be effective for the AWESOM, an ASRG is likely to be preferable. These are more than four times as efficient as RTGs, each one capable of producing 130 W of electricity from 500 W of heat generated from 1.2 kg of PuO_2 (plutonium dioxide). With the same quantity of PuO_2 as Curiosity (4.8 kg), which would require four ASRGs, the AWESOM could produce 520 W of electrical energy from 2000 W of heat. Further analysis is required to determine exactly how many ASRGs will be sufficient for water mining.

The ASRG will function as a source of both heat and electrical energy, with the electrical energy being partially utilised to generate microwave radiation. The heat and microwaves will both be directed at the ice, causing it to sublime, and releasing H_2O from the regolith without the need for digging. As described by Dr Robert Zubrin (Larson & Pranke 1999), the robot will have a flexible skirt brushing the dirt to contain the released water vapour beneath the robot. The moist air will then be sucked up through a dust filter using a fan, and the water extracted from the Martian air via adsorption into zeolite 3A in the same manner as in the WAVAR.

Keeping water in a liquid phase within the AWESOM's tank will require heat and pressure, because of the low temperatures and pressures at the Martian surface. The tank will therefore be pressurised, with heat provided by the ASRG. Heat from the ASRG can also keep the robot's electronics sufficiently warm. The hose for transferring water from the AWESOM robot to the MAV's tank will also need to be heated, although this will be done electrically.

The AWESOM will return to the MAV to unload at the end of each sol, sparing it the energy cost of keeping the water in liquid phase overnight.

The AWESOM robot will have a simple robotic arm for connecting the end of the hose into a socket at the base of the MAV. It will also include a pump to transfer the water. Since electricity for microwaves and motors will not be required while the water is being transferred, it will be wholly available to power the robotic arm, pump, and heating of the hose during this process.

The robot will use radio communications with the MAV to keep track of what ground it has already covered, infra-red proximity obstacle detection sensors like those found on cars and vacuum robots, and a neutron moisture meter to measure the water concentration in the regolith. AWESOM will also feature cameras to send images back to Earth, which will assist with remote operation and debugging of the robot, should that be necessary. The robot will be designed to operate fully autonomously.

The density of dry Martian soil is about 1400 kg/m^3. If the average water concentration of the regolith is 7% by mass, and approximately 80% of this can be extracted from the top decimetre of regolith, the amount of water that can be obtained from each square metre of ground can be estimated as follows:

1400 kg/m^3 * 7% * 80% * 0.1 m = 7.84 kg/m^2

8.5.4. Water Strategy

An AWESOM robot similar to that sent with the MAV could be included with the SHAB, which would produce abundant water for crew requirements. However, this should not be necessary because:

1. Enough water can be collected more easily from the atmosphere.

2. The AWESOM robot from the MAV can be moved to the SHAB when the crew arrive.

The proposed water strategy is therefore as follows:

1. Bring no H_2 or H_2O from Earth. This significantly reduces the landed mass of the SHAB, and avoids the problems inherent in transportation and long-term storage of hydrogen.

2. Develop a WAVAR unit capable of collecting at least 2.05 kg of water per sol, and ideally at least 1.5-2 times this much, from the Martian atmosphere. This unit will be included with the SHAB and integrated with the ECLSS and ISAP systems (see In Situ Air Production), and capable of reliably collecting sufficient water for crew requirements during the surface element. The SHAB will include a high-efficiency (95%+) WRS, and storage capacity for approximately 10-20 tonnes of water. (Some or all of the water will be stored around the ceiling and walls to provide additional

radiation protection.)

3. On arrival at the IMRS, two crew members drive the CAMPER to the MAV, collect the AWESOM robot and bring it back to the SHAB. At this point the ISPP process will be complete, the MAV's propellant tanks will be full, and it will no longer need the AWESOM (assuming zero propellant boil-off or leakage during the surface mission). The SHAB will have a similar external water intake valve as the MAV, and the AWESOM robot can be programmed for collecting water from around the SHAB. This will provide abundant additional water for the crew, and function as a backup to the WAVAR, while affording the engineers an opportunity to study the robot's performance and improve its design. Assuming a rate of water production of around 29 kg/sol (see In Situ Propellant Production), the AWESOM robot could produce up to an additional 15 tonnes of water during the surface element. If the AWESOM cannot be retrieved from the MAV, or fails for some reason, the water collected by the WAVAR will be sufficient.

4. When Alfa Crew depart, there should be enough water left over in the tanks to support Bravo Mission. However, both the WAVAR and the AWESOM are able to operate autonomously and will continue to collect water for the SHAB during the 8 months before Bravo Crew arrive. If both are fully operational during this period, perhaps 5-10 tonnes of additional water can be collected during this time. If one or both fail, it won't matter as long as there's at least 1.2 tonnes in the tanks to support the minimum requirements of the mission. Bravo Crew may be able to repair the ISWP equipment when they arrive.

5. When Bravo Crew arrive they can collect the second AWESOM robot from the second MAV, and bring it to the SHAB where it can replace the first AWESOM. If the first AWESOM is still operational, both can be used together, one at each SHAB. Otherwise, it can be repurposed, or used as a source of spare parts. Its ASRG could be used to augment SHAB or CAMPER power.

As more ISWP assets are accumulated at the base, water will be available to astronauts in increasing abundance, and as the water is put to use for food production, the base will gradually become self-sufficient.

8.6. Food

The DRA details food requirements for a crew of six on a long-stay human mission to Mars. If IMRS uses the same food system as the DRA then the food requirements will be basically the same, because the crew size and approximate durations of both the space and surface element are the same.

However, with new innovations in food technology it may be possible to substantially decrease the mass, volume, energy and time requirements of the food system, in addition to significantly reducing waste.

8.6.1. Powdered Food

New powdered food products such as Soylent (Rhinehart 2013) are emerging, which claim to be nutritionally complete and may potentially carry a wide range of advantages, most of which are applicable to space missions. They require minimal time to prepare, as they only need to be mixed with water and a small amount of oil, require minimal cleanup, produce no waste (at least in the form of inedible biomass), have an extended shelf life, require minimal packaging and do not require refrigeration or cooking. There is also anecdotal evidence that subsisting primarily on such foods is beneficial for health and improves mental clarity, due to the low toxicity and absence of unnecessary calories and ingredients. By ensuring that all the body's nutritional needs are met in an optimised way, this type of food may be healthier than a typical human diet (not that this implies it's the healthiest possible human diet).

Powdered food seems extremely well-suited to long-duration space missions, and is even sometimes called "space food". It's lightweight and compact, can cater for the crew's nutritional requirements in an efficient way, and could potentially optimise their physical and mental functioning. The elimination of refrigeration and cooking equipment greatly reduces associated mass, volume and energy requirements.

The first human missions to Mars will need to be optimised for mass and energy as much as possible in order to make them affordable and achievable. Every kilogram and kilojoule saved in terms of food, packaging, waste, and food storage and preparation equipment means a lighter and cheaper mission, reducing cost and increasing viability.

In addition, the ease with which meals can be prepared saves considerable time. The astronauts may be so busy, especially once on Mars, that they may appreciate not having to devote so much time to food. To mix and consume a shake a couple of times per sol could save an hour or more in time spent preparing and consuming food, and cleaning up afterwards.

It's important to remember that this is an historic exploration mission, not a holiday. In this regard, analog studies do not reflect reality. Although life at the MARSes are undoubtedly busy, the stress level is low, the feeling is generally relaxed, and there is ample free time. Astronauts on Mars, however, will have a tremendous amount of work to do while conforming with strict rationing and safety regulations. Naturally there will be time for rest and recreation so the crew remain happy and healthy for the duration of the mission, but planners will be under pressure to achieve the greatest possible ROI, which means the crew will be given plenty to do, especially while on Mars. They may appreciate meals that are very quick to prepare and consume, as well as nutritious.

Formulas and flavours

Evidence of a substantial market for convenient and healthy food of this type has become apparent, suggesting that investment in this type of food technology will increase. The likely result will be numerous innovations, including improved formulas and flavours emerging during the next 1-2 decades; which is to say, before the first HMMs. A formula suitable for astronauts spending time in reduced gravity environments could be developed, perhaps higher in certain key nutrients such as calcium and protein, or even including bisphosphonates (a drug that prevents bone loss).

The flavour of Soylent is neutral. A wide variety of powdered flavour additives are available to be mixed in, including chocolate, vanilla, caramel, coffee, tea, various fruits, peanut butter, chilli, cheese, bacon and many others. A range can be included with the astronauts' food supply, allowing them to be creative and mix their own favourite combinations, and to vary these as desired without compromising their nutrition or investing significant time.

Reduced waste

Waste can be significantly reduced by using powdered food. For example, rather than using meal-sized plastic containers, powdered food can be stored in large drums, which can be repurposed when empty. This is in stark contrast to conventional space food, which generates approximately 300-500 grams of waste per person per day. For a 910-day mission with a crew of 6, this is over 2 tonnes of waste. Wasted food and water is also likely to be much less. With conventional space food, food packaging waste includes adhered food and food preparation water, which is typically not recovered. However, with powdered food, portion size is easy to control, a single serve is likely to be fully consumed, and the water content of adhered food can be recovered by rinsing the shaker and emptying the liquid into the WRS.

The optimised nutritional profile of powdered food means highly efficient utility by the body, and thus reduced mass and volume of solid waste. This correspondingly reduces the energy required to process the waste, the volume required to store it, and the amount of water lost in the recycling process. In addition, because the whole crew would be eating basically the same food for the entire mission, the chemical profile of waste products would be highly predictable, which may suggest opportunities for more efficient or useful methods of processing or recycling waste. Studying the metabolic waste products produced by powdered food will generate ideas for progressively improving the formula and making it yet more efficient, and producing even less waste.

Importance of shared meals

Research at the ISS and MARSes has shown that the experience of spending time with other crew members in familiar domestic situations such as preparing meals and eating together provides important psychological benefits and is effective for building relationships and morale. Food is something we all have in common, and meal times stimulate team communication, camaraderie, humour and fun. It's therefore a legitimate argument that providing the crew only with powdered food for 2.5 years could largely deprive them of these experiences, potentially exacerbating feelings of boredom or frustration, or even contributing to mental health issues.

Despite the proposed advantages of powdered food, it is therefore important not to underestimate the value of being able to prepare and eat "normal" meals. Angelo Vermeulen, a Belgian artist and scientist, recently led a 4-month NASA study at HI-SEAS about cooking and eating on Mars:

> The food that astronauts currently eat is pretty good, but it's all pre-prepared. It's add-water-and-heat, and you have your meal. But even those meals, even when they try to make variations, after a couple of months people get tired of that, and so they start to eat less. As a consequence they might also perform less, and jeopardize the mission.

> For example, in the Mars-500 experiment — an isolation study of 500 days near Moscow, a collaboration between Europe and Russia — food became the item that people constantly talked about. Food is absolutely crucial to the psychology of your crew, and you need to handle that carefully.

> One of the solutions could be to allow the crew to cook. Because cooking empowers you over your food. You can make endless variations, and there's an interesting bonus: it improves social cohesion. You talk about food, you share food. It's a basic human thing. The reason that space agencies have been holding it off are twofold. First of all, current human space exploration is done in microgravity conditions — like in the ISS — and as such cooking has hardly been possible. One needs a good deal of gravity to cook meals. In HI-SEAS we're talking about simulating life on the surface of Mars, not about traveling to Mars. And since there's a decent amount of gravity on Mars (38% of Earth's gravity), you can do your regular cooking.

Cooking in microgravity in the THAB may be difficult, but in the SHAB experimentation with cooking would represent valuable research in addition to fulfilling an important social function.

Food system for IMRS

If a group of astronaut candidates was asked if they would still want to go to Mars if it meant living only on powdered food for 2.5 years, there's little doubt they would accept that condition. The sociopsychological benefits of shared meals could be generated in some other way, such as morning meetings over tea/coffee, or group social activities such as movies and games.

However, the psychological state of the crew is one of the biggest risks to the mission, and research shows that food is an important part of that.

It may be that a compromise is therefore optimal, with powdered food being consumed one or two meals per sol, but with a social crew meal at other times, plus snacks of favourite conventional food. This balance would still generate an appreciable reduction in mass, volume, energy, time and waste, while allowing opportunities for the astronauts to be creative and prepare meals for each other, and to have an enjoyable social experience.

Therefore, the IMRS missions are currently envisaged as providing 50% "normal" food and 50% powdered food. This will produce some of the benefits of using powdered food, without the crew going crazy from food boredom or lack of social interaction.

The conventional food utilised during the space element will be the traditional "add water and heat" pre-prepared meals, as these are designed for the microgravity environment. However, for the surface element, normal food will be provided in the form of ingredients that can be combined and cooked in the SHAB, thus providing a pleasurable, social, interactive experience once per day, at the evening meal. Breakfast and lunch can be powdered food, which is quick and easy, and will give the crew the necessary energy and nutrition to do their work. Snacks during the day could also be normal food, such as dried fruits, chocolate, nuts, etc.

Naturally the crew will also have access to abundant supplies of the finest organic coffee, tea and cocoa that Earth can produce.

8.6.2. Food Mass Estimates

For the purpose of this exercise, only the dry mass of food is considered, not the mass of packaging or food preparation equipment.

Conventional space food

The standard food requirement is 0.634 kg/p-sol (ed. Hanford 2004). As with water consumption, EVA increases food requirements by an additional amount, which is 0.029 kg/p-h (0.715 kg/p-sol). The same estimates for time spent on EVA as used in calculating water requirements must also be used: an average of 80 msol/sol during the space element, and an average of 183 msol/sol during the surface element.

Powdered food

The mass required may be estimated from the standard serving size of Soylent, which is 166 grams including the oil blend. This represents one third of daily nutritional requirements for a typical adult, and should therefore be increased by 2.75% to calculate the nominal quantity for a sol. The requirement is therefore:

0.166 kg * 1.0275 * 3 = 0.512 kg/sol.

Adjustments to caloric and nutrient requirements may be necessary to support the physical stresses of the mission, which could affect the mass of daily requirements. However, this figure will serve for a first-order analysis.

Assuming the same ratio applies between IVA food requirement and EVA additional food requirement as for conventional food (about 113%), the additional powdered food required for EVA will be approximately 0.577 kg/sol.

Space element

In addition to water, the DRA includes contingency food in the THAB:

The food that is carried aboard the transit habitat includes transit consumables that are needed for the round-trip journey plus contingency consumables that are required to maintain the crew should all or part of the surface mission be aborted and the crew forced to return to the orbiting MTV, which would then function as an orbital "safe haven" until the TEI window opens. Any remaining contingency food remaining on board the crewed MTV would be jettisoned prior to the TEI burn to return home.

Therefore, the food requirement for the space element must cover the full mission duration of 885 sols, not just the 350 sols of the space element. Considering the plan to provide the crew with 50% conventional food, for a crew of 6, the mass of conventional food required during the space element is:

(0.634 kg/p-sol + (0.08 * 0.715 kg/p-sol)) * 50% = 0.346 kg/p-sol

Mass of powdered food:

(0.512 kg/p-sol + (0.08 * 0.577 kg/p-sol)) * 50% = 0.279 kg/p-sol

Total per sol:

(0.346 kg/p-sol + 0.279 kg/p-sol) * 6 p = **3.75 kg/sol**

The minimum amount of food that *Adeona* needs to depart Earth with is therefore:

3.75 kg/sol * 885 sol = **3319 kg**

The food requirement for the return trip will be:

3.75 kg/sol * 175 sol = **656 kg**

Surface element

During the surface element, the required mass of conventional space food is:

(0.634 kg/p-sol + (0.23 * 0.715 kg/p-sol)) * 50% = 0.399 kg/p-sol

The required mass of powdered food is:

(0.512 kg/p-sol + (0.23 * 0.577 kg/p-sol)) * 50% = 0.322 kg/p-sol

The total food mass required for the surface element is therefore:

(0.399 kg/p-sol + 0.322 kg/p-sol) * 6 p * 535 sol = **2314kg**

8.6.3. In Situ Food Production

Explorers and settlers of Mars will certainly want to grow their own food, for food security, convenience, independence, variety, and long-term sustainability of settlements. Even from the very first mission, it's likely that astronauts will experiment with food production.

Crews in at least the first few human missions to Mars will not be able to rely on locally produced food, as food is mission critical, and considerable research will be necessary to learn how to grow it on Mars. In addition, food production requires volume, energy, equipment, nutrients and water, which won't all be available during those early missions. Therefore, all the food required for the space and surface elements of the mission will be transported from Earth, and this will continue to be the case until the necessary agricultural technologies and processes have been developed for Mars, and the necessary resources are available, to reliably produce sufficient food. Even once food production is underway on Mars, it may still be years before a nutritionally complete variety of crops is being locally grown.

Researchers at Wageningen University have experimented with growing crops in Martian soil simulant with good results (Wamelink 2013). Nonetheless, Martian explorers and settlers face a range of challenges related to ISFP:

1. Growing plants requires equipment, such as pipes, pots, hoses, trays, racks, lights, filters, tools and more.

2. Growing plants requires volume. Whether growing the plants in Martian dirt, or using a hydroponic or aeroponic system, a significant volume will be required to produce food, even for a crew of six. Providing this volume would necessitate a larger SHAB, or a dedicated greenhouse. In addition, production of nutritionally-complete food for the crew would require growing a wide range of different plants, and it may be difficult to to grow a large number of different crops in a restricted volume.

3. Growing plants requires water. Whether transported from Earth or obtained locally, water will be a precious resource during early missions. ISWP assets will need to be operational at the base and producing abundant water to support food production. (Note that the amount of water required to grow plants can be greatly reduced by using aeroponics.)

4. Growing plants requires environment control. If the greenhouse is separate from the SHAB it will require its own ECLSS, as sufficient heat will be required to prevent plants from freezing, and CO_2 and humidity levels must be controlled to optimise growth rates. This ECLSS will require additional energy. Even if crops are grown in the SHAB, or in a greenhouse connected to the SHAB, thus leveraging its ECLSS, a larger unit with higher energy requirements will be needed.

5. The lower gravity reduces the rate of exchange of gases such as O_2 and CO_2, which causes plants to grow slower and evaporate less water. Greenhouses may require a higher atmospheric pressure or elevated CO_2 levels or temperature to compensate.

6. Growing and eating plants produces waste in the form of inedible biomass. This could be composted with the crew's solid waste into fertiliser. However, compost is only really useful if the plants are grown in soil, and aeroponic systems may be preferred during the early stages of exploration and settlement of Mars. Simply dumping biomass on the surface of Mars may violate planetary protection protocols. Biomass may need to be incinerated.

7. Growing plants requires light. Solar intensity on Mars is only about half that of Earth, which is sufficient for photosynthesis, but may also contribute to slow growth rates. Sunlight within greenhouses may be supplemented with additional lighting. To control energy requirements, LEDs (Light-Emitting Diode) that produces photons of the specific frequencies used by plants (blue, red and a little infra-red) may be utilised. Plants can even be grown indoors using only this type of lighting.

High-tech indoor vegetable farm in Singapore (Credit: Panasonic)

Clearly Martian agricultural processes will be very different from those on Earth. In spite of these challenges, ISFP is a fundamental enabling capability that needs to be developed for human settlement of Mars. Human missions should therefore incorporate experimentation into food production while at Mars. The DRA also considers ISFP to be a key objective for human habitability of Mars.

One advantage unique to Mars is that, because the sol is only marginally longer than a day, it should be easy for Earthian plant species to adapt to Mars's diurnal cycle.

Growing methods

On Earth, the name for growing crops in dirt is "geoponics"; hence, on Mars the technically-correct term is "**areoponics**", although perhaps this term will not enjoy common usage due to potential obvious confusion with "aeroponics". Martian soil contains all the nutrients necessary to support plant growth, and could be made yet more amenable to plants via the addition of organic matter and micro-organisms such as nitrogen-fixing bacteria and mycorrhizal fungi.

One of the advantages of using dirt to grow plants is that it's available in abundance and easily obtained. Another advantage is that metabolic waste from the crew, and food and vegetable scraps, can be composted and mixed with Martian dirt to produce fertile soil.

The disadvantages of growing food in dirt are:

- More area is needed compared with other methods.

- Water requirements are comparatively high.

- Separate greenhouses are necessary because dirt isn't wanted in the SHAB (see <u>Airlocks</u>).

Hydroponics may be preferable option, which is how Mars One are proposing to grow food in their habitats (see below). Hydroponics obviously requires water, however, this can be recycled through the system with high efficiency. One question is whether the plant nutrients would be brought from Earth, somehow extracted from the Martian dirt, or both.

Aeroponics, in which plants are suspended in air and nutrients are delivered to their roots via a nutrient-filled mist, may be a promising option for the IMRS due to the low water requirements. Lettuce, strawberries and many other types of plants can be grown aeroponically. However, separate greenhouses are required for areoponics also, as mist in the SHAB could affect electronics and crew health, and produce mould.

An advantage of aeroponics is that the plants have full access to the O_2 and CO_2 in the air. The O_2 acts as a purifier and helps to keep plants healthy, and CO_2 levels in the greenhouse atmosphere can be optimised to maximum plant growth rates.

Aquaponics is a combination of aquaculture and hydroponics, in which fish tanks are connected to a hydroponics setup in a mutually beneficial symbiotic relationship. Water containing fish waste, which is high in plant nutrients, is channeled from fish tanks through the hydroponic system to feed the plants. Conversely, plant cuttings and food scraps are used to feed the fish.

Aquaponics may become extremely popular on Mars, as fish will provide an abundant and welcome source of protein and essential fatty acids. If suitable tanks can be incorporated into a greenhouse design, it may be possible to transport fingerlings to Mars, for example salmon, barramundi or tilapia, and grow them at the IMRS.

Other animal-based foods

Other animals could gradually be introduced to the IMRS for the purpose of food production:

1. Stingless bees could be utilised in greenhouses for plant propagation, and would also produce honey.

2. Worms may be introduced for processing mineral grains, fertilising soil, and feeding fish. They can also be eaten by humans and are a good source of protein.

3. Insects such as crickets are another potential protein source that could be grown relatively easy. These are planned for Mars One.

4. Chickens are low-maintenance, would be a source of meat as well as eggs, are highly effective at processing food scraps into fertiliser, and make great pets.

Mars One and PlantLab

The Mars One mission requires the astronauts to grow their own food on Mars. If successful, their efforts will certainly be instructive in future ISFP attempts. Their intention is to grow plants inside the habitat modules using a hydroponics system similar to PlantLab. PlantLab is a highly efficient indoor crop production system developed in the Netherlands, with low volume, mass and energy requirements.

Following research conducted by NASA, plants will be grown under LED lighting that only produces frequencies used by plants, thus optimising energy requirements. Natural sunlight will not be used at all. Crops will be arranged in vertical racks to reduce space requirements, and the intention is to produce enough food for 12 astronauts with just 50 m^2 of growing area.

Fittonia plants growing under LED lighting at PlantLab (Credit: Peter Dejong/AP)

9. Mars Transfer Vehicle

In this section, several design ideas for MTVs are reviewed and compared in order to produce a basic design concept for *Adeona*.

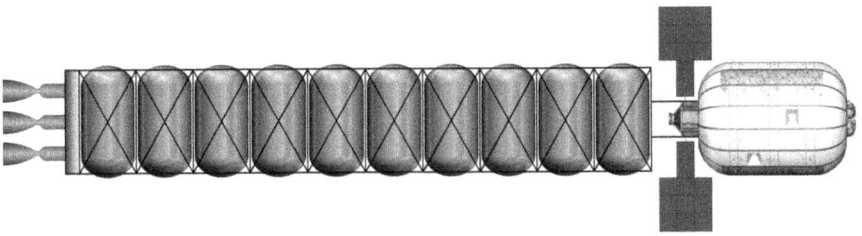

Mars Transfer Vehicle concept

The Mars Transfer Vehicle (known as *Adeona*) carries the crew to Mars. In Blue Dragon, as in the DRA, the MTV also brings the crew back to Earth. This is in contrast to Mars Direct, in which the crew launch from Mars and fly back to Earth in the ERV. *Adeona* is constructed on orbit, only used in space, and does not land on any planetary surface.

An MTV, or at least, the process of getting the crew to Mars and back, is arguably the most challenging component of HMMs to design, and surely the most expensive to build when considering the total cost of development, materials, propellant, launches and construction. For these reasons, it's the one element that tends to generate the most innovation, as mission designers look for creative ways to transport humans to Mars more cheaply, quickly and safely.

The reason why getting to Mars is still so expensive is largely related to our current level of technology:

- The only currently available form of propulsion that could potentially be used for a trip to Mars is chemical propulsion. However, an MTV requires hundreds of tonnes of propellant to reach Mars, irrespective of the choice of propellant.

- Travelling to Mars via chemical propulsion is slow, with one-way trip times of around 5-8 months. This increases cost due to the amount of supplies that must be carried on board, plus measures necessary to protect the crew from radiation while in space, such as shielding.

- Rockets are large, complicated and expensive, and, as yet, neither reusable nor mass-produced. Thus launch costs are still very high.

- We don't currently have any SHLLVs flying, which increases the number of launches needed to construct and build an MTV, and making on-orbit assembly a likely requirement. Both of these inflate the cost.

- All the propellant and rockets are currently on Earth, which means propellant must be transported from the surface of Earth to Earth orbit. This is expensive.

- Transfer and storage of cryogenic propellant in space incur losses due to boil-off, which increases the initial quantity of propellant to be launched from Earth, and therefore also the cost.

- Current technology does not provide a way to keep astronauts 100% healthy in the interplanetary environment, which means trip times must be kept as short as possible. This means bigger propellant burns when entering and exiting orbit at both Earth and Mars, further increasing propellant requirements.

- Currently available materials that would be strong enough for an MTV are still very heavy, which increases launch costs. It also increases the quantity of propellant required to move the vehicle, which *also* increases launch costs.

- We've never done it before, which means most of the necessary hardware needs to be developed from scratch, and extensively modelled and tested. Engineers, tools, and manufacturing and testing facilities are also expensive.

These reasons suggest a number of areas for technological research and innovation that will reduce the cost of space travel:

- Faster/better/cheaper methods of space transportation. Ideas that have been, or are currently in development (or at least discussion), include nuclear thermal rockets, ion thrusters, field propulsion, space elevators, antimatter rockets, wormholes, warp drive and more.

- Reusable rockets, such as those currently in development at SpaceX. This should reduce launch costs significantly, since each rocket can be used for multiple launches instead of just one.

- Technologies and systems for on-orbit transfer and long-term space storage of cryogenic propellant.

- More advanced materials and manufacturing processes that will result in improved radiation shielding, lighter spacecraft components, higher performance rocket engines and zero boil-off propellant tanks.

- Technologies for creating artificial gravity fields.

- Asteroid and lunar mining to create propellant sources in shallower gravity wells (if chemical propulsion is still relevant in the same timeframe).

- Space manufacturing to produce spacecraft components and other hardware in shallower gravity wells.

Requirements for IMRS

A minimum of two MTVs are required for the IMRS plan. This is because a full mission takes about 30 months, but the Earth-Mars synodic period, which defines when crews can be launched to Mars most economically, is 26 months. Therefore, *Adeona* will still be 4 months away from Earth when it's time to send the second crew to Mars. (The second MTV departs Earth approximately 2 months before Mars opposition, whereas *Adeona* arrives back at Earth approximately 2 months after it.)

Advantages of having two MTVs:

1. Missions can be launched every 26 months instead of every 52.

2. One serves as a backup for the other.

A generous 22 months are available between an MTV's return to Earth orbit and its next departure for Mars, in which time it can be refurbished and refilled it with propellant, and training missions conducted.

9.1. The Transit Habitat

Adeona is basically comprised of a propulsion stack plus the transit habitat, a.k.a. the "THAB". Its main function is to propel the THAB and the crew to Mars and back. In the IMRS plan the THAB is comprised of a B330. Once fully inflated, whether this is done on Earth or orbit, the THAB will be fitted out with fixtures, fittings and equipment such as:

*   communications and computing hardware

*   furniture

*   gym equipment

*   food storage and preparation equipment

*   water storage

*   crew cabins

*   laboratory equipment and scientific experiments

As discussed earlier, it is likely to be preferable to fit out the B330 on Earth rather than launch it deflated and fit it out on orbit. This strategy doesn't fully take advantage of the benefits of an inflatable habitat, which is to reduce launch costs by launching the module in a deflated state, thus permitting the use of a smaller rocket. However, sending people to Mars is a considerably more ambitious mission than constructing on-orbit space stations, and budget should be available to use a larger vehicle capable of launching a completed THAB. Fit-out of the THAB on Earth by expert engineers and tradespeople with access to the best tools will produce a much better result than sending up one or more small crews to attempt such tasks as installing furniture and computer systems in microgravity.

Because the THABs are reused, a minimum of only two need to be built and launched anyway, which justifies the higher cost of launching them fully fitted out.

The THAB will be designed for a microgravity environment. The internal volume may be divided into rooms or sections by adding walls or "floors", either parallel or perpendicular to the habitat's axis.

Private cabins/bedrooms

The THAB is large enough that each crew member can be allocated their own private space where they can sleep, read, receive personal communications, meditate, listen to music, and activities of this nature.

Gym/medical

As discussed in <u>Health and Fitness</u>, it is necessary to provide the crew with exercise equipment for both cardiovascular and strength training. This equipment will be located in a room that may also double as a medical bay, complete with the equipment necessary for blood work and other health examinations.

Control/meeting room

A dedicated area will be necessary for morning meetings, with computers for inspecting the spacecraft condition and other work, and communicating with MCC on Earth. Here may also be located the ship's proxy server, which maintains a local cache of research material, multimedia, websites, games and other software, and acts as an internet gateway, providing the crew with online services such as email and social networking.

Food

A dedicated area where food can be accessed, prepared and consumed without the risk of it floating around the ship and affecting experiments or instruments.

Laboratory equipment and scientific experiments

Half the crew will be scientists, and they will wish to take advantage of the unique opportunity to conduct research in deep space. The crew members will design and prepare experiments that align with their particular interests. A particularly useful experiment will be growing food in space, such as strawberries, to add variety and pleasure to the crew's diet.

9.2. Reusing the MTV

Mars Direct, the DRA and Blue Dragon each take a different approach when it comes to reusing the MTV:

1. In Mars Direct the crew fly out to Mars and land on Mars in the SHAB. They launch from Mars and return to Earth in the ERV, which performs a direct entry into Earth's atmosphere.

2. In the DRA, the crew travel out to Mars and back in the MTV, with an Orion capsule docked. As the MTV approaches Earth, the crew transfer to the capsule, undock, and perform direct entry into Earth's atmosphere, splashing down in the ocean. The MTV continues past Earth to enter orbit around the Sun.

3. In Blue Dragon, the crew fly out and back in the same vehicle, as in the DRA. However, on return to Earth, *Adeona* is captured into HEEO, then descends to LEO. A Falcon 9 with the Earth Descent Capsule is then launched from Earth, it docks with the THAB, and the crew descend to Earth. *Adeona* can then be reused in subsequent missions.

The advantage of returning to Earth via direct entry is that no propellant is required for an EOI manoeuvre, which means less is needed for all previous manoeuvres. This results in a significant reduction IMLEO. However, there are distinct disadvantages:

- There is little room for error. In the case of Mars Direct, the propellant burn must be exact in order for the ERV to enter Earth's atmosphere at precisely the right moment and angle so that it won't burn up, skip out, or crash. In the DRA, for the same reasons, the Orion capsule must be released at precisely the right moment and angle as the MTV flies by Earth.

- A fast, direct-entry EDL is risky, dangerous and stressful for the crew, subjecting them to very high g-forces.

- The return vehicle is lost; to run another mission will require building another.

Another disadvantage of the Mars Direct approach is that because the ERV is launched from Mars it must be made very light. The living volume must therefore be minimised, which could be unpleasant for the 6-month return trip.

A disadvantage of the DRA approach is that the Orion capsule for Earth descent must be carried all the way to Mars and back, adding a nontrivial mass penalty.

The cost of the IMRS strategy is additional propellant for the EOI burn, which must be carried to Mars and back. This additional mass, when propagated back through the architecture, results in a nontrivial increase in IMLEO and therefore also the initial cost of the MTV. However, there are numerous benefits that justify this approach:

- *Adeona*, an expensive piece of hardware, will not be wasted, can be used repeatedly, making it cheaper, quicker and easier to run each mission. This reduces space debris and spares the considerable cost of building a new MTV. It will be much easier to get approval from sponsors for another HMM when there's already a perfectly good vehicle ready and waiting in LEO. Like an airplane, in principle the vehicle will only need a maintenance inspection and refilling with propellant to make it ready for the next mission. The THAB can be cleaned, and *Adeona* repaired and/or upgraded as required.

- The mass of additional propellant is partially offset by the mass of *Newton* (the Earth Descent Capsule), which doesn't need to be taken all the way to Mars and back.

- The deceleration forces experienced by the crew during Earth EDL are significantly less when descending from LEO than when approaching by direct entry. This is safer, and considerably less risky and stressful for the crew. Descending to Earth in this way will be reasonably routine for the crew, having done the same thing several times before, both from the ISS and from *Adeona* during training.

- Once the THAB is on Earth orbit there will be plenty of opportunities to launch a capsule to pick up the crew.

- *Newton* can land on solid ground back at the launch pad, eliminating the need for a water recovery and simplifying quarantine procedures.

- As discussed below, reuse of the THAB takes advantage of the tweaks, refinements and homely touches made by previous crews, as with FMARS and MDRS.

- Training missions can be conducted to the MTV immediately without having to wait for financing and construction of a new one.

Each MTV will have their own unique name and associated character, like ships of the sea. The word "astronaut" means "star sailor", after all. Crews, not to mention the millions of fans, will be reluctant to see them simply discarded.

9.2.1. No Place Like Home

During the 6-month outbound trip, the crew will make the THAB their home. They'll humanise the space, putting up posters, sticky notes, hooks, photographs and knick-knacks, with each person making little tweaks to the environment to suit their individual behaviours and preferences, and to improve safety and comfort. They'll develop routines that work with the spacecraft and develop protocols for co-habitation in the relatively confined volume. They'll learn the various quirks of the vehicle, and find a whole range of optimisations and workarounds. The vehicle's server will have all their favourite music, movies, books, games and websites already downloaded. They will fix and improve things — an improvised washing line here, a cable-tie or pencil-holder there.

For example, in the picture below, note the photographs, pens, and other odds and ends affixed to the walls behind the astronauts' heads on the ISS:

Astronauts Michael Fincke and Sandra Magnus, and cosmonaut Yury Lonchakov. (Credit: NASA)

This is one of the many things people do to personalise a space, and it has significant psychological value. This is another reason why it will be better to use the same vehicle for both outbound and return legs, and also to retain the THAB for multiple missions.

For the spacecraft's designers, decorations and personal touches will be a secondary consideration compared with the primary function of the hardware: to keep the crew alive and carry them safely to their destination. The designers of the THAB interior are unlikely to consider everything six unique individuals will need or want in order to create a comfortable living and working space for 1 year. The crew will fill these gaps themselves during their period of habitation.

By using the same ship for both outbound and return trips, there will be a sense of familiarity, almost home-coming, when they re-enter *Adeona* for the return journey. Everything will already be prepared, by each crew member, for themselves; they can return to the comfort of their individual sleeping cabins, prepared on the way out to Mars. This will have a positive effect on morale and will eliminate the time cost of personalising a separate return vehicle in the same way as the outbound vehicle.

In the event that any members of the crew are coming back from Mars sick or injured, having a familiar environment will even be important.

When the same THAB is inhabited by new Mars crews, this will also have important psychological benefits via a powerful feeling of connection with past crews, and a feeling of confidence in the tried-and-tested vehicle and habitat. New crews can draw inspiration from, and add to, the various marks and refinements made by previous ones, as has happened over the years in FMARS and MDRS.

The THAB is a valuable mission artefact that one or more crews of "marstronauts" have made their home for a year. If a new one is created for each mission, the only sources of ideas for improving the design would be astronaut reports and memories, and video footage. A physical artefact is significantly more useful. The original designers will not think of everything, hard as they may try; every mission to Mars and back will reveal new ideas for how the MTV needs to be improved.

9.2.2. Save Money

There's no doubt that getting to Mars and back is the most expensive part of the program; at least for now. There's simply no getting around this until space technology improves. This provides a strong incentive to come up with an innovative, lean vehicle design within the constraints of the available resources; and to make the vehicle reusable. If the intention is truly to run numerous missions to Mars, then reusability is essential for cost-effectiveness.

Reusing the MTV will provide an opportunity to perform maintenance and make repairs or improvements to the engines, propellant system, solar panels, radiator arrays, computer and life support systems, or any other components, depending on available finances and priorities. Additional inflatable modules could be added so that the vehicle can support larger crews, along with lengthening of the structure to provide additional propellant pod bays and engines.

To illustrate the cost benefit, let's imagine an MTV that costs $1 billion to construct, but is completely discarded at the end of each mission. A new one is required each time; therefore, for three missions, the cost of building MTVs is $3 billion. Now let's estimate that the cost of building a *reusable* MTV is $1.5 billion, due to the additional propellant. However, to refurbish the MTV and refill it with propellant for subsequent missions costs only $150 million. The total cost for three missions is therefore $1.8 billion — a saving of 40%. And there's no reason to stop at three missions.

A minimum of two MTVs are required because of the mission timetables. However, the same mathematics applies, and becomes more apparent when a minimum program of 10 missions is assumed. Using the same rough estimates, constructing a new MTV for each of 10 missions would cost $10 billion. However, with two reusable MTVs, each being used for five missions, would only cost $4.2 billion (2 * $1.5 billion + 8 * $150 million) — a saving of 58%.

Clearly this could mean the difference between a small number of missions, as described in the DRA, or a large number of missions, which is the goal for IMRS. A small number of missions carries the risk of an Apollo-like result, i.e. a decade of awe-inspiring achievement and glory followed by decades of tooling around in LEO. In contrast, an ongoing series of missions focused on accumulating hardware and infrastructure on Mars will lead to settlement and the establishment of a new branch of civilisation. Reusable MTVs may cost more, but their value to the human race are inestimable.

It must be considered that, while government terms of office are typically 3-5 years, the IMRS program will span at least 20-30 years. Because reusability greatly reduces the cost of future missions, the likelihood that those missions will be approved by new governments is much higher.

When *Adeona* is replaced by a new vessel, she can be repurposed. For example, it will be possible to run training missions to the spacecraft, operate it as a space station or laboratory, use it for human missions to other destinations in the Solar System, rent or sell it to private enterprise, or even provide tourist experiences or preserve it as a museum. After all, it will be a tremendously historic artefact.

A new propulsion mode such as electric, ion, warp, or field propulsion may need to be developed before rapidly reusable MTVs are practical. Alternately, a nearer-term solution may yet be realised with the benefit of nanostructured materials, which will produce spacecraft hardware so much lower in mass that a fully reusable MTV using chemical propulsion could become more practical.

The ultimate goal is an SSTO (Single Stage To Orbit) spacecraft that can rapidly fly from the surface of Earth to the surface of Mars, refill with propellant from local Martian resources, and fly back to the surface of Earth, all without discarding any major pieces of the vehicle and without even needing to wait for an optimal launch window at either end. Once such a vehicle exists, a process of settlement can develop in earnest.

9.3. Propulsion

9.3.1. Chemical vs. Nuclear

The DRA examined three basic options for in-space propulsion: chemical, electric and NTR (Nuclear Thermal Rocket). Of these, NTR was preferred.

Blue Dragon, however, favours chemical propulsion. Chemical propulsion is well-proven, has a much higher TRL, and is much cheaper, safer and better understood than NTRs or any other form of propulsion. A mission can be designed around chemical propulsion without much guesswork about cost and performance. The salient features of the IMRS program — safety, low cost, and minimal development time — all point towards chemical propulsion.

The primary benefit of NTRs is their relatively high I_{sp}, of the order of 1000 seconds. This is more than twice LOX/LH2 (liquid oxygen and liquid hydrogen) and almost three times that of methalox, which means much less propellant is required. However, NTRs would be much more expensive to develop, as their TRL is still quite low.

Although some development work on NTRs has been completed, none have ever actually flown. The NERVA (Nuclear Engine for Rocket Vehicle Application) program flight tested several NTR components; however, that program closed in 1972, which is over four decades ago at the time of writing. Project Timberwind and the Space Thermal Nuclear Propulsion program conducted further research into NTRs, however, failed to reach their objective of flight testing an NTR upper stage, and the program was closed in 1994 (two decades ago at the time of writing). Some in the space community believe that NTRs will never be funded to full development, and that constructing Mars missions based on NTRs creates an artificial obstacle. Others believe that, with new technology, much better NTRs can now be developed at much lower cost.

An NTR is a minimally-shielded nuclear fission reactor, which, most would agree, is rather a dangerous piece of equipment. In the event of failure, the entire spacecraft including the crew could be exposed to high levels of radiation far more harmful than a solar particle event.

The use of NTRs in near-term Mars missions may technically be possible, but for other reasons seems unlikely. They would take much longer and cost more to develop, would be difficult to get approved and to test, are more dangerous, and are unnecessary. In addition, most NTRs use hydrogen as the working fluid, and, as discussed below, hydrogen has its own set of problems.

Exotic propulsion systems that reduce the time and cost involved in getting to Mars will indeed be valuable to future missions, and essential for future expansion into the Solar System, Milky Way and Universe. However, just as sailing ships were perfectly adequate for worldwide European expansion without needing to wait for steam ships or aeroplanes to be invented, chemical propulsion is available now and is perfectly capable of transporting crew and cargo safely to Mars. A 6-month flight to Mars is comparable to an 18th century ocean voyage from England to Australia, except considerably more comfortable, as the crew will be living in environment-controlled conditions with plenty of amusements, food and water, and always able to communicate with friends and families.

In fact, there's a valuable benefit in taking the slow road to Mars, which is the free return trajectory. If the crew decide not to capture into Mars orbit, they can slingshot around the planet and be back at Earth 2 years later, without minimal or no propulsive burns. This is a valuable abort option to have, especially for the first few human missions. If the flight to Mars is much quicker, a free return abort option will not be available.

Other exotic propulsion methods such as ion propulsion, VASIMR (VAriable Specific Impulse Magnetoplasma Rocket), plasma thrusters, SEP, NEP, field propulsion, etc., are not currently under consideration for use in Blue Dragon, as they are speculative or still in development, are not needed, and their inclusion may introduce delays. However, when new propulsions technologies become available that are superior in terms of cost, safety and performance, they should certainly be considered.

9.3.2. LOX/LH2

The chemical propulsion option considered in the DRA for the MTV is LOX/LH2. This bipropellant is often favoured for two main reasons:

- Highest specific impulse for liquid bipropellant (~450 seconds).

- Availability of existing engines.

- Non-toxic.

- Unlike hydrocarbon fuels, combustion of LH2 doesn't produce any pollution — only water. (Although, for in-space propulsion this is not an issue.)

- Available throughout the Solar System, and, indeed, the Universe.

However, for *Adeona*, or indeed any Mars mission, LOX/LH2 is probably *not* the ideal choice as it has a number of nontrivial drawbacks.

1. LH2 is notoriously difficult to store for long periods, especially in the vacuum of space. Because of the very small molecule size, it tends to leak away, or "boils off". Even with a boil-off rate of only 1%, it would still be necessary for *Adeona* arrive at Mars with a surplus of at least 20% of the fuel needed for TEI. Note, however, that zero boil-off tanks and densified liquid hydrogen have been developed to address this problem.

2. LH2 has very low density (71 kg/m^3 compared with 424 kg/m^3 for LCH4 or 810 kg/m^3 for RP-1), and therefore requires larger storage tanks

compared with other fuels. The tank size becomes yet larger when accounting for the boil-off margin. Larger tanks are obviously heavier, and thus more expensive to launch.

3. LH2 is highly cryogenic, with a boiling point of just 20 K, and therefore requires active cooling. This additional power requirement means additional mass for solar panels or other power system hardware, thereby further increasing IMLEO and launch costs. The need for active cooling means that LH2 is not generally considered space storable. This makes it unsuitable for *Adeona*, which must wait 1.5 years on Mars orbit.

4. Because LH2 must be stored at such low temperatures, its tank must be kept separate from the LOX tank, otherwise the LOX could freeze. This increases the mass of tankage and other parts of the propellant system.

5. LH2 causes metal to become brittle, which means associated engines and tankage require advanced metallurgy. This also drives up the cost of the vehicle.

6. LH2 is more expensive than either RP-1 or LCH4.

7. LH2 is highly explosive and difficult to work with, and can cause invisible high-temperature fires.

8. Because of its low temperature, LH2 fuel lines can only be purged with expensive helium.

These drawbacks arguably outweigh any advantages of LH2 attributable to its high I_{sp}.

Advancements such as the new composite cryotanks developed by Boeing and NASA may produce strong, lightweight LH2 tanks that greatly reduce boil-off rates, possibly to zero when combined with MLI (Multi-Layer Insulation). However, even if this technology becomes available, considering the other disadvantages of hydrogen there is probably a better choice for MTV propellant.

9.3.3. Methalox

Instead of LOX/LH2, a more suitable option is almost certainly methalox. As settlement of Mars and the Solar System progresses, methalox engines may become increasingly common, at least while chemical propulsion remains economical.

Methalox offers a range of important advantages as a bipropellant:

- **Availability, low cost, and potential for ISPP.** Methane and oxygen can easily be obtained or manufactured from resources abundantly available throughout the Universe. On Earth, methane can be obtained directly from natural gas, of which it is the major constituent, and oxygen can be obtained directly from the atmosphere. On Mars, both can be manufactured from readily available CO_2 and H_2O. C-type asteroids, which comprise about 75% of known asteroids, contain abundant carbon and water from which methalox can be manufactured. On Titan it rains methane; there are lakes of the stuff. Long range exploration and settlement of the Solar System will be greatly enhanced by vehicles that can refill with propellant from resources discovered along the way.

- **Commonality of hardware elements.** Methalox is favoured for Mars ascent due to the potential for manufacturing it from local Martian resources. It could therefore be advantageous to use the same propellant for other manoeuvres, such as Earth ascent, Mars descent, the MTV, and surface vehicles. Methalox is suitable for launch as well as interplanetary and surface vehicles. If the same propellant is used in multiple applications, redundancy becomes automatically built into the knowledge base. Engines will be similar, parts counts and cost will be reduced, reliability increased, and propellant can be transferred from one application to another, thus improving propellant security. Propulsion engineers working on different vehicles using similar engines will be able to share expertise and innovations, resulting in greater synergy and problem-solving capability, and more rapid evolution of methalox engine technology. If methalox engines become more common, and are studied by more people, new ways to optimise and improve them will emerge.

- **Non-toxic.** This, in addition to other factors, reduces ground-handling costs and makes it considerably easier to get permits for engine testing.

- **Cryotank design.** By controlling pressures, LOX and LCH4 can be stored at a common temperature. In fact, the LOX tank can be nested inside the LCH4 tank, with the tanks thermally coupled so that only the LOX tank requires cryocooling (Grayson, Hand & Cady 2009). This reduces the total volume required for tankage within the vehicle, and the mass of tanks, cryocooling systems, insulation and superstructure.

- **Pump design.** The volume ratio between is LOX and LCH4 is about 1.3, or nearly equal. The similar temperatures and volume flow rates mean that centrifugal pumps can be used with almost identical requirements, and could perhaps be run off a single spindle. This simplifies pump design, further reducing mass.

- **Better than RP-1.** Methane has various other advantages over RP-1:

  - cheaper;

  - slightly better specific impulse;

  - cleaner, not being subject to coking or polymerisation;

  - can be run less O_2-rich, which is easier on the pumps.

- **Better than LH2.** Methane has various advantages over hydrogen:

  - cheaper;

  - only requires passive cooling, and much less insulation;

  - has a much higher density, and thus requires smaller tanks;

  - has a higher boiling point, which means fuel lines can be purged with ordinary gaseous nitrogen.

One feature of LCH4 is that it requires an ignition source to initiate combustion. This can be viewed as another advantage over LH2 the sense that LCH4 can be kept in close proximity to LOX without risk of exploding. It may also be considered a disadvantage because ignition can be difficult at low temperatures. However, it should be well within the capabilities of 21st century engineers to design methalox engines capable of reliable ignition, including restart capability, at Martian and deep space temperatures.

9.4. Shedding Mass

In the DRA, the mass of the MTV is reduced *en route* by jettisoning pieces of the spacecraft that are no longer needed, such as propellant tanks and engines. This is worthwhile because, the lighter the vehicle, the less propellant is required to move it. Any reduction in mass at any point in the journey will translate to an appreciable reduction in propellant requirements, which, in turn, can mean millions of dollars saved.

9.4.1. Cruise Stages

A spacecraft's cruise stages comprise the engines, propellant tanks and associated hardware required for interplanetary propulsion, often attached to a long, rigid truss-like structure. Some interplanetary spacecraft designs include only one cruise stage, whereas others have multiple. As with rockets, one benefit of having multiple stages is that, as each fulfils its purpose, it's discarded in order to shed mass and thereby decrease the propellant required for the next manoeuvre. The net effect is an overall reduction in IMLEO and therefore cost. However, IMLEO is not the only consideration and there are a variety of trade-offs that need to be considered.

For the journey to Mars and back, four major manoeuvres are necessary:

Approx. sol	Manoeuvre	Description	Approx. Δv (m/s)
0	Trans Mars Injection (TMI)	From LEO to an outbound MTO	3560
175	Mars Orbit Insertion (MOI)	From outbound MTO to HEMO[1]	900
710	Trans Earth Injection (TEI)	From HEMO to an inbound MTO	900
885	Earth Orbit Insertion (EOI)	From inbound MTO to LEO	3560

[1] Highly Elliptical Mars Orbit

When using chemical propulsion, because of the low I_{sp} it's usually considered more efficient for the vehicle to have multiple stages, each with their own engines and propellant, and for each stage to be discarded after its dedicated manoeuvre. The image below shows the chemical propulsion version of the MTV proposed in the DRA. (The vehicle has no EOI stage, as it doesn't capture into Earth orbit.)

Each engine in this design is an RL10B-2, which is a LOX/LH2 rocket engine developed by Aerojet Rocketdyne.

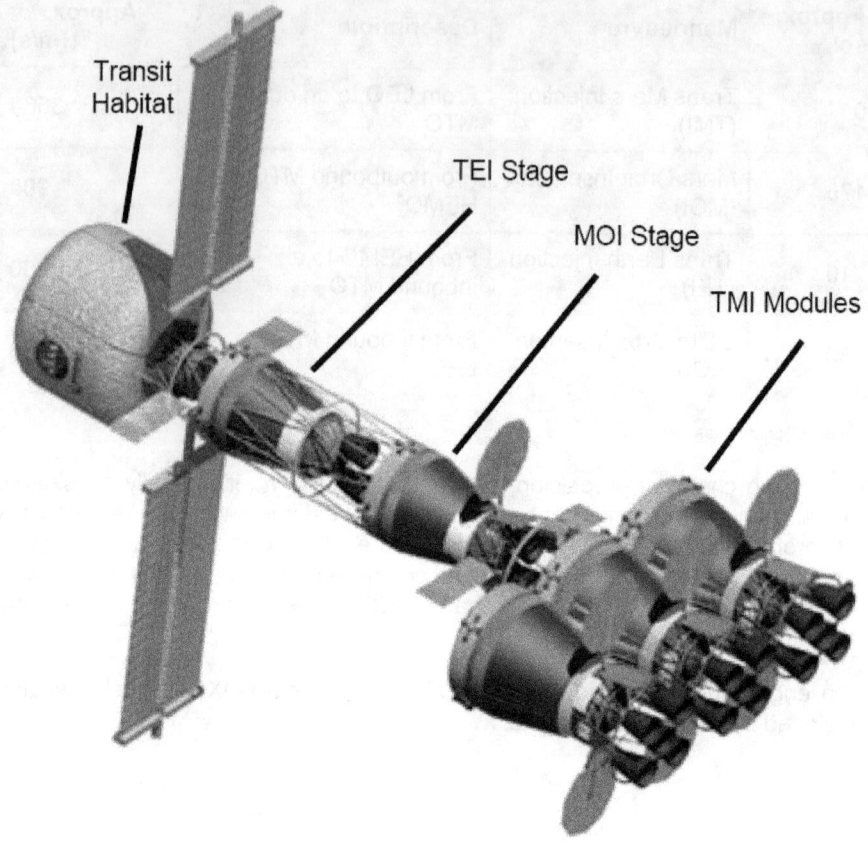

Chemical-propulsion MTV from NASA's DRA (Credit: NASA)

9.4.2. Drop tanks

Rather than discarding stages to reduce mass, another approach is to discard propellant tanks that are empty and no longer required. This reduces the mass of the vehicle *en route*, without the loss of engines or superstructure.

This is the approach taken in the DRA with the NTR design. One large LH2 fuel tank (the "drop tank") is discarded after the major TMI burn in order to reduce mass, as shown in the bat chart below.

The International Mars Research Station

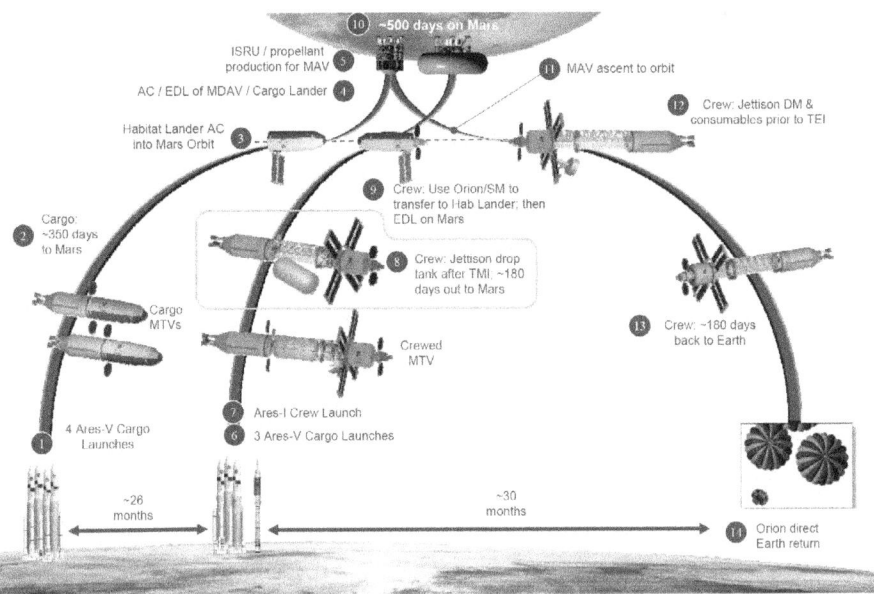

Bat chart for the DRA showing the NTR MTV and drop tank jettison (Credit: NASA)

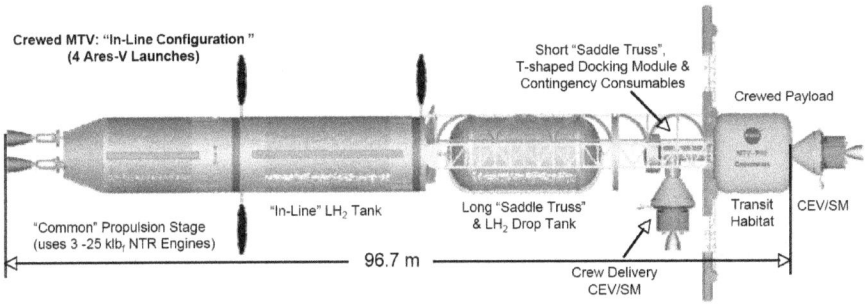

From the DRA: MTV concept with NTR propulsion and drop tank (Credit: NASA)

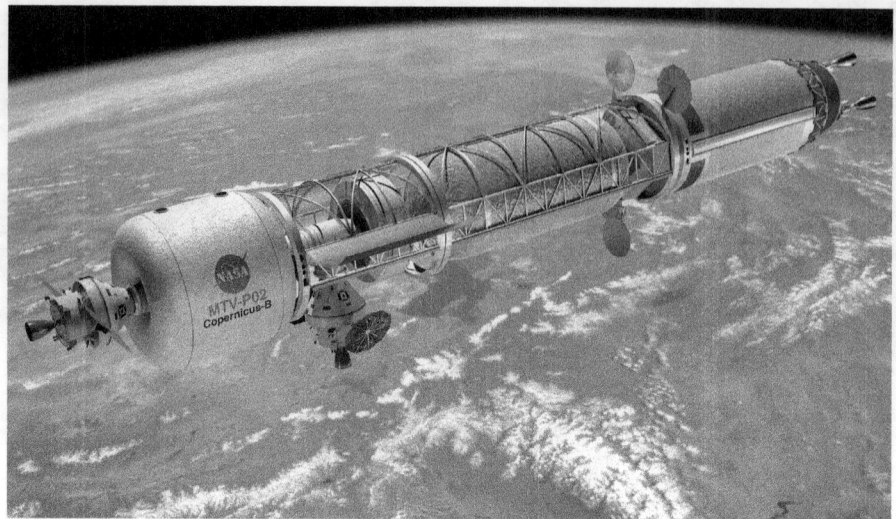

From the DRA: An NTR MTV waiting in LEO (Credit: NASA)

The advantage of drop tanks for a reusable MTV is that replacing them once back in Earth orbit would be inexpensive compared with replacing superstructure, engines or entire stages. This use of drop tanks may therefore also be a good idea for *Adeona*, and may also suggest a simple and effective mechanism for initial filling and subsequent refilling of the vehicle with propellant.

Propellant must be transported to Earth orbit in tanks anyway. Considering the difficult in on-orbit transfer of cryogenic propellant, it may be simpler and more efficient to deliver entire tanks and plug them directly into the MTV.

9.4.3. Expendable Methalox Propellant Pods

Below is shown a concept for an "Expendable Methalox Propellant Pod" (EMPP), designed to fit inside the fairing of a Falcon Heavy:

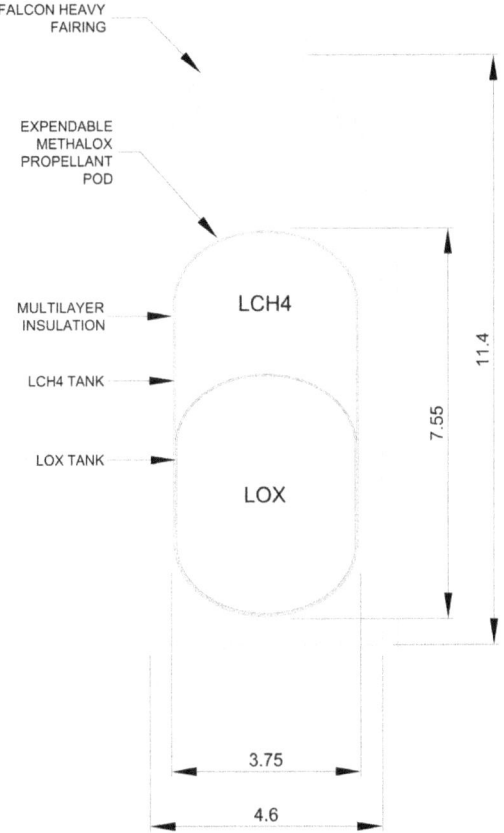

FALCON HEAVY
FAIRING

EXPENDABLE
METHALOX
PROPELLANT
POD

MULTILAYER
INSULATION

LCH4

LCH4 TANK

LOX TANK

LOX

11.4

7.55

3.75

4.6

EMPP in a Falcon Heavy fairing

The purpose of the EMPPs is to enable a number of key benefits:

1. Each is jettisoned when empty, reducing *Adeona*'s mass by about 1.5 tonnes.

2. They obviate the need for in-space cryogenic propellant transfer. The entire tank is transferred from the delivery vehicle (Falcon Heavy) to *Adeona*.

3. They enable the vehicle to be refilled with propellant at reasonably low cost using reusable rockets.

This diagram shows:

- The inner LOX tank.

- The outer LCH4 tank.

- MLI (Multilayer Insulation) surrounding the tanks.

The EMPP is actually a small spacecraft. Not shown in the diagram is a simple GNC and OMS (Orbital Manoeuvring System) necessary for docking the pods. The battery-powered electronics would only need to operate for a few minutes. Each pod will locate *Adeona* via radio, orient itself automatically, and guide itself into its allocated pod bay, where external sockets on the pod will plug into *Adeona*'s propellant lines. The OMS thrusters would be fuelled by a small quantity of methalox from the main tanks.

Nested Composite Cryotanks

The propellant pod design shown above combines two recent innovations in propellant tank technology: composite cryotanks, and thermally-coupled nested methalox tanks.

Cryotanks made of composite materials, such as those developed recently by Boeing and NASA, will be ideal for the MTV as well as the MAV. These mass about 70% of traditional metal tanks, with a specific weight of approximately 8.3 kg/m^2.

Composite cryotank developed by Boeing and NASA (Credit: Boeing, NASA)

Engineers at Boeing have also invented a thermally coupled LOX/LCH4 tank (Grayson, Hand & Cady 2009), which substantially reduces the overall mass of cryocooling systems and insulation. This is made possible because, unlike LH2, LCH4 can be stored at the same temperature as LOX.

In the optimal configuration, an inner LOX tank is contained entirely within an outer LCH4 tank:

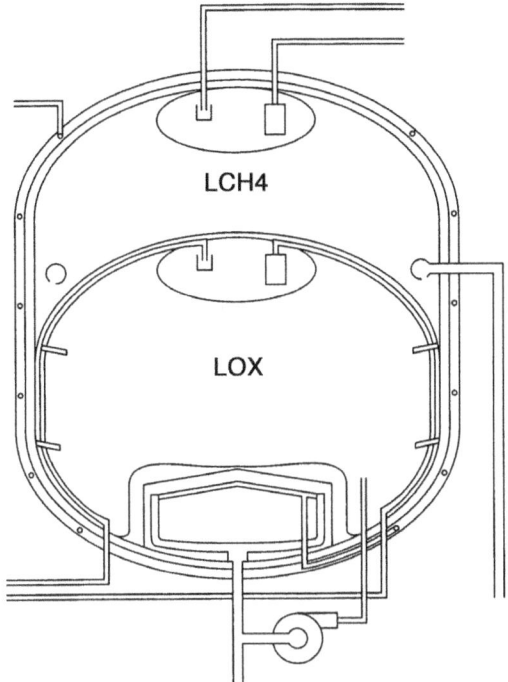

Thermally coupled LOX/LCH4 propellant tanks (Credit: Grayson, Hand & Cady 2009)

Only the LOX tank is cryocooled. The LCH4 is cooled passively by heat transfer through the common tank wall. By controlling pressures, both tanks can be kept at a common temperature of 91 K.

Further mass savings are gained by the optimised geometry of the tanks, which reduces the mass of insulation and superstructure.

9.4.4. Food, Water and Waste

As described in the Resources section, *Adeona* must depart Earth with a minimum supply of 1564 kg of water plus 3319 kg of food (dry mass), which is a total of 4883 kg. However, prior to TEI propulsion burn for the return home, any surplus contingency food and water, plus any unrecyclable waste generated during the mission so far, will be jettisoned in order to reduce mass as much as possible.

Minimum supplies needed for the return journey are 485 kg of water plus 656 kg of food, for a total of 1141 kg. The jettisoned mass is therefore:

4883 kg - 1141 kg = 3742 kg

This is the sum total of any remaining contingency food and water, plus dry metabolic waste products (extracted from urine, faeces and sweat), generated during the space element up to that point.

In addition, at least a tonne of additional solid waste, including food packaging, toilet paper, sanitary napkins, medical waste, inedible biomass, etc., will also have been generated and can be jettisoned prior to TEI. However, this is not included in the basic calculations presented here.

It's worth noting that propellant can be optimised even further by jettisoning accumulated waste before MOI and EOI.

9.5. Methalox Engines

Adeona's methalox engines are currently envisaged as having a thrust of approximately 1.75 MN.

While it would be congruent to use a COTS engine, no methalox rocket engines are currently in active use, and none of this size have ever been developed (although concepts have been proposed). Therefore, a new engine will need to be developed.

SpaceX is currently developing a large new methalox engine called "Raptor", which will power the second and upper stage of the Falcon 9 and Falcon Heavy rockets, and its 10 metre-diameter Mars Colonial Transport vehicle. Based on SpaceX's track record, this will probably be quite a good engine with an efficient design, and at least partly 3D printed, like SpaceX's SuperDraco engine. It's I_{sp} is expected to be 380 s, which is high for a methalox engine. However, with a vacuum thrust of 8.2 MN, this engine is too large for the MTV. A lower thrust per engine is preferable in order to control g-forces experienced by the crew.

Blue Origin, in collaboration with ULA, are also developing a new methalox engine, with a thrust of about 2.4 MN. While this is a better match, it may still be too large for the MTV design described here, although could perhaps be suitable if the engine can be throttled. If so, this would result in significant savings in development costs, plus the fact that only three engines per MTV would be required.

The estimated I_{sp} of the new engine is also 380 s, which should be achievable. Assuming a conservative TWR (Thrust-to-Weight Ratio) of 100, each engine will mass about 1835 kg each. Five are required, for a total engine mass of around 9175 kg.

The advantage of using an odd number of engines is so that any number can be used depending on the thrust required, while still keeping the direction of thrust collinear with the vehicle's axis by firing engines in a balanced way:

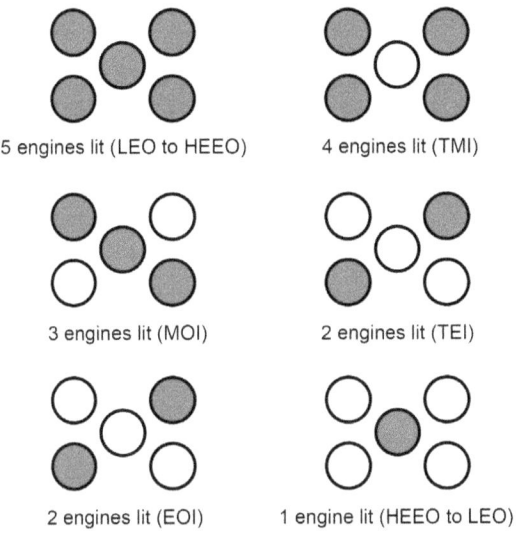

5 engines lit (LEO to HEEO) 4 engines lit (TMI)

3 engines lit (MOI) 2 engines lit (TEI)

2 engines lit (EOI) 1 engine lit (HEEO to LEO)

Patterns for engine firings to maintain balanced thrust

The number of engines used for each manoeuvre is determined by *Adeona*'s maximum acceleration, which is kept below 4 g's for crew comfort and health, and so they can still reach the controls.

9.6. *Adeona* Design

The design outlined here is simplistic, does not represent a complete or optimised design, and is offered purely to communicate a few ideas and hopefully to stimulate further creativity and discussion from sufficiently motivated allies.

Some conservative assumptions are made about the technological developments of the coming 1-2 decades:

- It will be cheaper, easier and quicker to develop better chemical rocket engines, and to launch a large quantity of propellant to LEO, that to develop entirely new propulsion technologies.

- COTS components such as inflatable habitats with adequate radiation shielding (e.g. the B330) and reusable rockets (e.g. the Falcon Heavy) will be available.

- An HLLV (Heavy Lift Launch Vehicle) — currently assumed to be the SLS, although other candidates may emerge — will be available for launching sections of *Adeona*.

- Cryogenic propellant tanks will be manufactured from composite materials rather than metals.

- Structural materials will become significantly lighter.

The concept for *Adeona* has evolved from the following design goals:

- Almost fully reusable spacecraft. Only the propellant pods are jettisoned.

- Fly from LEO to HEMO and back.

- Chemical propulsion using methalox.

- As in the DRA, no aerobraking, aerocapture, artificial gravity, SEP tugs, propellant depots, Lagrange Points or gravity assists are utilised in the trip to Mars and back.

Adeona is comprised of:

- The THAB, with a docking port for a Dragon capsule.

- Solar panels and thermal radiators.

- GNC and communications equipment.

- RCS/OMS for orientation changes.

- A long structural truss-like section.

- 10 EMPPs (Expendable Methalox Propellant Pods).

- Five 1.75 MN restartable methalox rocket engines.

The International Mars Research Station

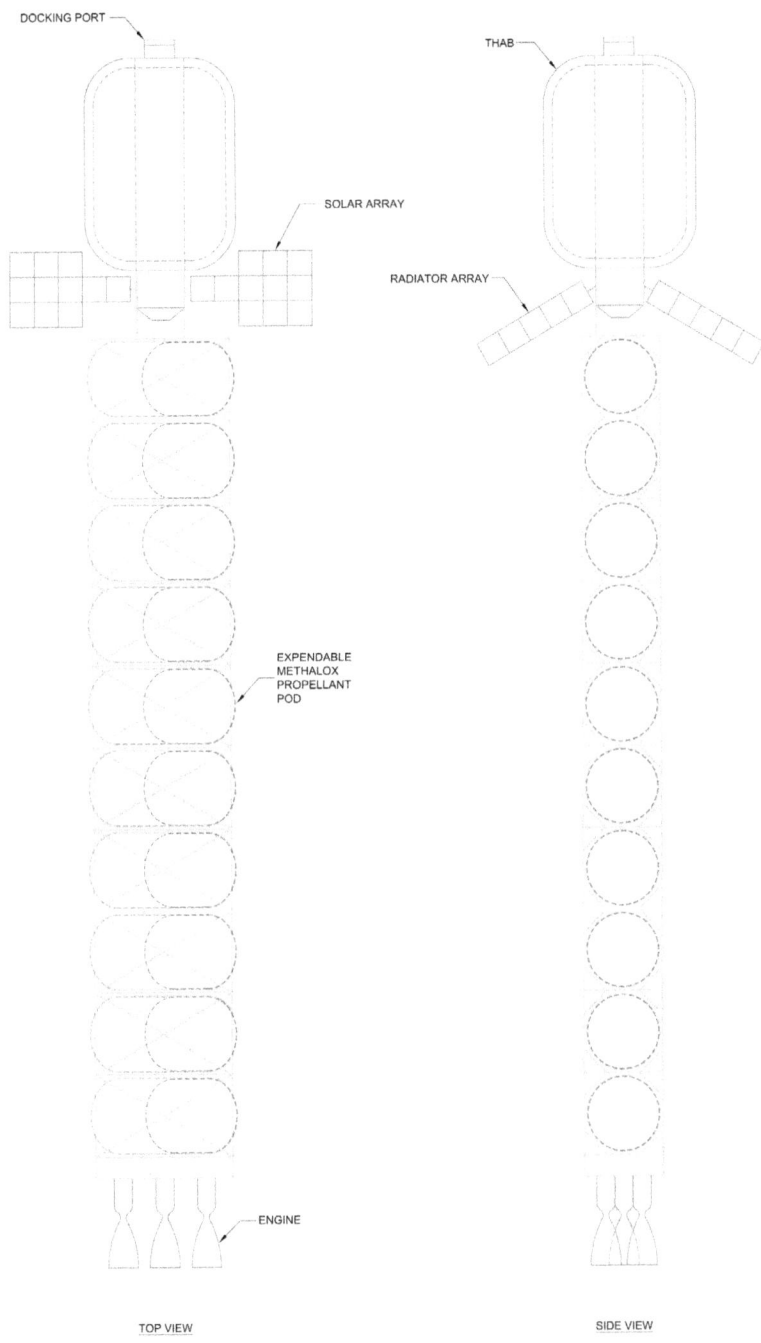

Adeona design concept - top and side views

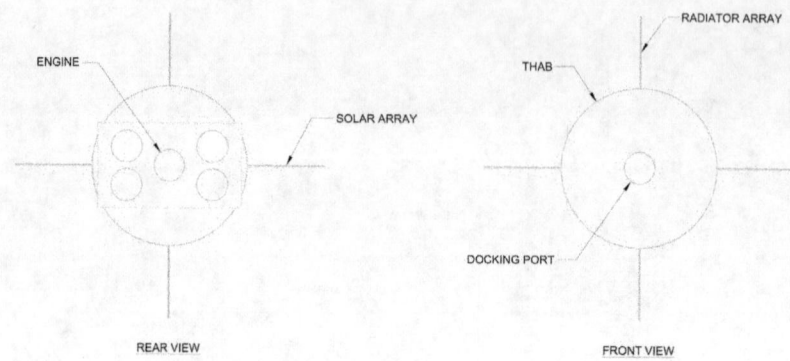

Adeona design concept - rear and front views

9.6.1. Trip Run-Through

A full run-through of the trip to Mars and back is described below to help explain the benefits of the EMPPs and how *Adeona*'s mass is optimised during the journey.

Climbing out of LEO

The first six EMPPs are used for the climb from LEO (nominally 400 km circular) to HEEO (nominally 400 km x 100,000 km). The climb is achieved with a series of six "perigee kicks", making best use of the Oberth effect.

One entire pod of methalox (about 51.5 tonnes) is burned each time *Adeona* passes through perigee, then the empty pod is jettisoned. This minimises the mass of the vehicle for each burn, thus maximising Δv and the altitude reached.

Apart from the fact that the EMPPs provide a much cheaper and more practical way to supply *Adeona* with propellant, this technique also efficiently reduces the total propellant requirement for TMI compared with doing one massive burn out of LEO.

The total Δv achieved during the climb is about 2.8 km/s, and if completed in six successive orbits, will take about 2.5 sols.

Climbing out of LEO

From Earth to Mars

Two manoeuvres are required to travel from HEEO to HEMO.

Adeona performs TMI from the HEEO, and flies to Mars along an elliptical orbit for 175 sols. When close to Mars, the MOI burn is performed, with *Adeona* capturing into HEMO with periareon at about 250 km altitude and apoareon at about 33,850 km altitude. This orbit has a period of 1 sol and is therefore often referred to as a 250 km x 1 sol orbit[*]. This particular orbit is also used by other Mars mission architectures, as it offers the benefits of being much easier to reach than LMO (Low Mars Orbit), and is synchronous with the base location, passing overhead once per sol.

After MOI, *Adeona* jettisons the 7th empty EMPP. The crew descend to Mars in the Blue Dragon capsule, further lightening the vehicle by about 5.2 tonnes.

Adeona then waits in this 250 km x 1 sol parking orbit for 535 sols while the crew do their thing on Mars.

[*] In the DRA, the parking orbit at Mars is 250 km x 33,793 km. The different apoaerion altitude is due to the inclination of the orbit and the fact that Mars' radius decreases slightly with increasing latitude. The important thing is not the precise altitude of apoaerion, but the 1-sol period.

From Mars to Earth

At the end of the surface mission, the crew ascend to *Adeona*'s parking orbit, and *Kepler* docks with the THAB. The crew transfer into the THAB, and *Kepler* is jettisoned. About 6.7 tonnes of contingency food and water is also jettisoned, being no longer needed. This makes the vehicle as light as possible prior to the TEI burn.

After TEI, the 8th propellant pod is jettisoned, and the crew enjoy a relaxing 6-month flight back to their home planet, satisfied in a job well done and thankful for the rest.

On arrival at Earth, *Adeona* performs EOI into HEEO, then jettisons the 9th propellant pod.

Back to LEO

The final manoeuvre is straight back into LEO. Although this is a large Δv manoeuvre (around 2.7 km/s), there's no need to "climb down into LEO" as such, because now the vehicle is comparatively very light, and only one propellant pod is needed.

Adeona parks back in the assembly orbit with one EMPP remaining, which may contain a few leftover tonnes of contingency propellant, depending on how accurate the various manoeuvres have been throughout the mission. As it cannot be refilled on orbit, this EMPP must also be jettisoned. Ideally it will be reclaimed and reused.

A Falcon 9 is launched from Earth with *Newton*, an empty Crew Dragon capsule, to rendezvous with *Adeona* and pick up the crew.

Prepare for the next mission

It has been about 30 months since *Adeona* left LEO. She now waits patiently in LEO for approximately 22 more months before her next departure.

During this time, several missions to *Adeona* are conducted, at least some of which involve EVA and full inspection of the vehicle by astronaut engineers. The THAB is cleaned and tidied.

Adeona is then refilled with propellant in the form of 10 fresh EMPPs, launched by reusable Falcon Heavy rockets, before receiving the next crew and setting off again for Mars.

9.6.2. Calculations

This section is included purely for people interested in the calculations underlying this design. Other readers may wish to skip ahead.

Δv quantities are based on this map, extracted from the Solar System Δv map created by Dragos Ilas. Values assume that propulsion burns occur at periapsis for best use of the Oberth effect:

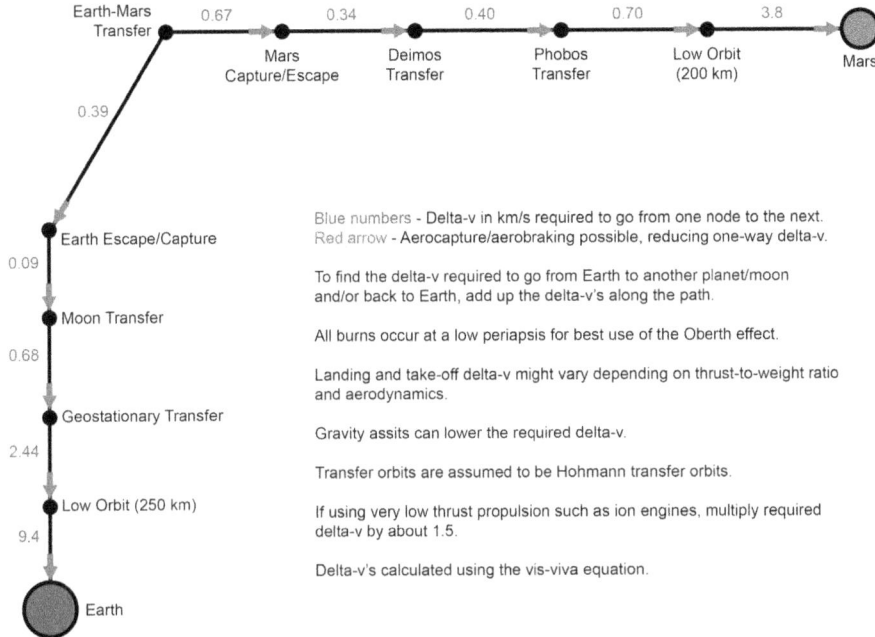

Earth-Mars Δv map

The values into and out of "Earth-Mars Transfer", a.k.a. Mars Transfer Orbit, will vary according to the speed of the orbit and thus the trip times, and these values may well be a little low. However, they should suffice for a first-order analysis.

Mass estimates follow — all values are in kilograms.

Adeona

The base mass of *Adeona*, excluding propellant, EMPPs and payload is estimated as follows:

Adeona base mass (kg)

THAB	22,000
Fittings	2000
Engines	9175
Structure, etc.	5500
Total	**38,675**

The mass of the 5 engines is determined from a nominal TWR of 100 for a 1.8 MN engine, which gives each engine a mass of around 1835 kg.

This structure mass may seem light, but advanced materials such as Ti-C composites are assumed. Miscellaneous masses such as navigation systems, RCS, propellant lines, pumps, thermal paint, etc., have not yet been accounted for as yet. Some elements will already be included in the THAB.

EMPP

A Falcon Heavy can deliver 53 tonnes to LEO. As the goal is to transport as much propellant to LEO as possible with each launch, the mass of a full EMPP matches this payload capacity.

The mass ratio of LOX to LCH4 in each pod is 3.5:1, which is the ideal stoichiometric ratio of methalox. The estimated mass breakdown is as follows:

EMPP component masses (kg)

LOX	40,040
LCH4	11,440
Total propellant	**51,480**
Tanks	1200
MLI	220
GNC/RCS/misc.	100
Total dry mass[*]	**1520**
Total wet mass[*]	**53,000**

[*]"Dry mass" refers to the mass of a spacecraft without propellant; "wet mass" refers to the mass of a spacecraft with propellant.

The estimated mass of the tanks is calculated from 8.3 kg/m^2, matching the recently developed composite cryotanks. The inner LOX tank surface area is about 56 m^2, and the outer LCH4 tank surface area is about 89 m^2.

The estimated mass of the MLI is calculated from an estimated 50 mm thick layer of double-aluminised Mylar with Dacron net spacer, with a bulk density of about 50 kg/m^3.

The 100 kg for GNC/RCS/miscellaneous is a guess.

Outbound payload

Outbound payload
masses (kg)

Crew	372
Marssuits	120
Baggage	120
Food & water	4883
Einstein	5200
Total	**10,695**

The world average mass of 62 kg for a human adult is used to estimate crew mass. Even though astronauts typically come from countries where adults have a higher-than-average mass, astronauts also tend to be fitter and lighter than the average.

According to Dava Newman, BioSuits are projected to have a mass of 20 kg each. In addition, each crew member is allocated an estimated 20 kg allowance for their individual kit bags. Admittedly, both these figures may turn out to be on the low side; however, note that these items must travel with the crew in capsules during Earth/Mars ascent and descent, and hence their mass and volume must be fairly restricted.

The 5200 kg mass of Einstein is based on an estimated 3300 kg for a Dragon V2 capsule without a trunk, plus 1900 kg of NTO/MMH propellant for landing on Mars (Grover et al. 2012).

Inbound payload

Inbound payload masses (kg)

Crew	372
Marssuits	120
Baggage	120
Food & water	1141
Samples	300
Total	**2053**

As discussed in <u>Food Mass Estimates</u>, approximately 3742 kg of surplus contingency food and water, and waste, is jettisoned prior to TEI.

Propellant burns

Initial state in LEO	
Spacecraft mass	38,675 kg
Payload mass	10,695 kg
Dry mass 10 EMPPs	15,200 kg
Propellant mass	514,800 kg
Total wet mass	**579,370 kg**
Burn 1: LEO (400 km) to EEO1 (400 km x 1781 km)	
Δv	347 m/s
Engines used	5
Total thrust	9 MN
Max acceleration	1.74 g
Propellant consumed	-51,480 kg
Drop EMPP 1	-1520 kg
Total wet mass	**526,370 kg**
Burn 2: EEO1 to EEO2 (400 km x 3779 km)	
Δv	384 m/s

Engines used	5
Total thrust	9 MN
Max acceleration	1.93 g
Propellant consumed	-51,480 kg
Drop EMPP 2	-1520 kg
Total wet mass	**473,370 kg**

Burn 3: EEO2 to EEO3 (400 km x 6933 km)

Δv	429 m/s
Engines used	5
Total thrust	9 MN
Max acceleration	2.18 g
Propellant consumed	-51,480 kg
Drop EMPP 3	-1520 kg
Total wet mass	**420,370 kg**

Burn 4: EEO3 to EEO4 (400 km x 12,685 km)

Δv	487 m/s
Engines used	5
Total thrust	9 MN
Max acceleration	2.49 g
Propellant consumed	-51,480 kg
Drop EMPP 4	-1520 kg
Total wet mass	**367,370 kg**

Burn 5: EEO4 to EEO5 (400 km x 26,600 km)

Δv	563 m/s
Engines used	5
Total thrust	9 MN
Max acceleration	2.91 g
Propellant consumed	-51,480 kg

Drop EMPP 5	-1520 kg
Total wet mass	**314,370 kg**
Burn 6: EEO5 to EEO6 (400 km x 110,574 km)	
Δv	666 m/s
Engines used	5
Total thrust	9 MN
Max acceleration	3.49 g
Propellant consumed	-51,480 kg
Drop EMPP 6	-1520 kg
Total wet mass	**261,370 kg**
Burn 7: EEO6 to MTO (TMI)	
Δv	685 m/s
Engines used	4
Total thrust	7.2 MN
Max acceleration	3.38 g
Propellant consumed	-43,874 kg
Total wet mass	**217,496 kg**
Burn 8: MTO to HEMO (250 km x 33,854 km) (MOI)	
Δv	900 m/s
Engines used	3
Total thrust	5.4 MN
Max acceleration	3.22 g
Propellant consumed	-46,666 kg
Drop EMPP 7	-1520 kg
Drop *Einstein*	-5200 kg
Drop surplus food and water, and waste	-3742 kg
Load samples	+300 kg
Total wet mass	**160,668 kg**

Burn 9: HEMO to MTO (TEI)

Δv	900 m/s
Engines used	2
Total thrust	3.6 MN
Max acceleration	2.91 g
Propellant consumed	-34,473 kg
Drop EMPP 8	-1520 kg
Total wet mass	**124,675 kg**

Burn 10: MTO to HEEO (EOI)

Δv	1004 m/s
Engines used	2
Total thrust	3.6 MN
Max acceleration	3.85 g
Propellant consumed	-29,445 kg
Drop EMPP 9	-1520 kg
Total wet mass	**93,709 kg**

Burn 11: HEEO to LEO

Δv	2556 m/s
Engines used	1
Total thrust	1.8 MN
Max acceleration	3.89 g
Propellant consumed	-46,514 kg
Total wet mass	**47,196 kg**

Final state in LEO

Spacecraft mass	38,675 kg
Payload mass	2053 kg
Dry mass 1 EMPP	1520 kg
Propellant mass	4948 kg

Total wet mass	47,196 kg

9.7. On-Orbit Assembly

Adeona is assembled in LEO at an "assembly orbit", which is nominally a circular orbit with an altitude of 400 km, i.e. basically the same as the ISS. This orbit is chosen mainly due to the extensive experience that has been gained by international space agencies operating at this orbit, including on-orbit assembly, and for proximity to the ISS in case it can provide support to the MTV, or in case it could be useful to connect *Adeona* to the ISS for training purposes or to aid with construction. However, as none of these things are strictly necessary, an alternative assembly orbit may be considered if there are advantages.

Adeona is launched as two separate sections, each of which fits inside the payload fairing of an SLS 105t Cargo. The solar panels and radiator arrays (not shown) are folded up to fit inside the fairing, or may be attached separately.

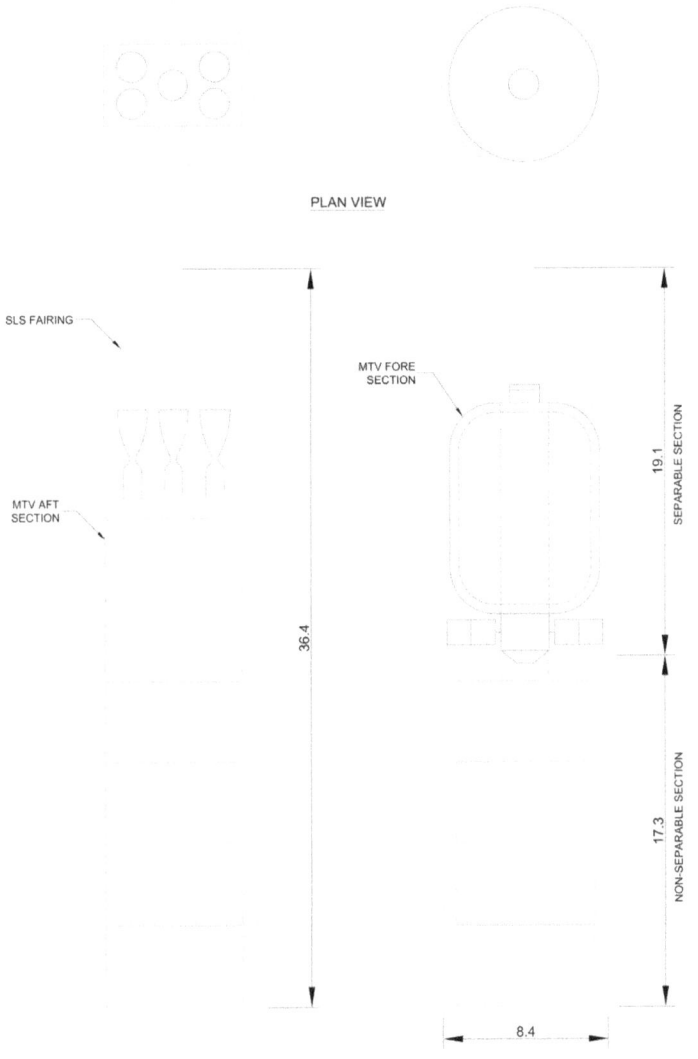

PLAN VIEW

MARS TRANSFER VEHICLE INSIDE SLS FAIRINGS

In Mars Direct, on-orbit assembly is presented as a drawback because of the additional cost and risk involved. However, without on-orbit assembly, the MTV must be much smaller, which would translate to less living volume, no potential for reuse of the vehicle, and perhaps a more expensive, or harder-to-develop propulsion system. On-orbit assembly enables the construction of a more capable, comfortable, and reusable vehicle, which can probably be available sooner and will cost less over the course of the IMRS program.

Although it adds some additional cost and complexity to the architecture, on-orbit assembly is hardly beyond current space engineering capabilities. The ISS was assembled in LEO, and was considerably more complex than *Adeona* will be. It's true that the Space Shuttle, which was a key element in ISS construction, will not be available, but recall that the Apollo LEM and command modules were docked on orbit without the benefit of the Space Shuttle.

The payload capacity of the SLS 105t is ample, since, without the methalox propellant pods, *Adeona*'s dry mass, excluding the EMPPs but including THAB fittings plus all food and water for the space element, is under 50 tonnes. In fact, it may be possible to launch each half with one full propellant pod already loaded.

As shown above, the payload fairing of the SLS 105t has a diameter of 8.4 metres and a total length of 36.4 metres, comprised of a 17.3-metre lower non-separable section and a 19.1-metre upper separable section. The 105t may be the smallest SLS with a cargo mode, but if an SLS 70t can be configured for this payload fairing then it might be a better choice, since it will be cheaper than the 105t, and its capacity of 70 tonnes to LEO is still way more than enough.

9.8. Reducing Propellant

The primary drawback of an MTV based on methalox is the large amount of propellant required to get to Mars and back, which makes the vehicle large and expensive. However, the total propellant requirement could potentially be reduced through a variety of strategies.

Aerobraking and aerocapture

It may be possible to use aerobraking at Mars and Earth to reduce *Adeona*'s velocity, rather than relying on propulsion alone.

Recall that:

- *Aerocapture* is when a spacecraft flies through the top of the atmosphere from an inbound transfer orbit, capturing it into an elliptical orbit.

- *Aerobraking* is when the spacecraft flies through the top of the atmosphere at periapsis of an elliptical orbit in order to reduce the eccentricity of the orbit; the opposite effect of a perigee burn.

The primary disadvantage of aerobraking and aerocapture is that the spacecraft requires some protection from the heat generated due to friction. As the spacecraft's velocity decreases, its kinetic energy is converted to heat. If the speed reduction is to be significant, then the heat generated will be proportionally significant.

Thermal protection can be in the form of a heat shield, or a coating of ablative material. An ablative material would almost certainly be non-renewable, which would affect the reusability of the spacecraft. In addition, it's difficult to imagine how a vehicle shaped like the MTV could be coated in such a material.

A heat shield has the advantage that it would also provide protection from atmospheric drag forces that could damage solar panels, radiators, antennas and other exposed equipment. However, for a spacecraft with a diameter of 6.7 metres (~15 metres if counting the full extent of the solar panels and thermal radiators) an inflatable heat shield would mass at least 5-10 tonnes. This additional mass would increase the propellant requirement, counteracting any potential savings achieved by using aerobraking.

Assuming, therefore, that *Adeona* does not carry a heat shield, she would not be able to go very deep into the atmosphere. She could potentially fly through the very tops of the atmosphere without thermal protection, and without overheating, but the effect on her speed would be minimal.

Aerocapture would therefore not be applicable to *Adeona*, because this must be achieved in a single manoeuvre. Aerobraking, however, can gradually lower an orbit over a long period by slowing the spacecraft down each time it passes through the atmosphere at periapsis. Because the change in velocity each orbit during aerobraking will be small, it could take months to lower the spacecraft's apoapsis to the desired altitude, which would seem to conflict with the goal of minimising the amount of time that the crew is in space. However, if a Dragon capsule can descend from a higher orbit without imposing excessive deceleration forces on the crew, they won't actually need to be onboard during this process.

Thus, on approach to Mars, *Adeona* could capture into a very highly elliptical orbit around Mars, and the crew could descend to Mars from this orbit. Then, during the 1.5 years that the crew is on Mars, *Adeona* could use aerobraking to gradually reduce her orbit to the target 250 km x 1 sol parking orbit. This strategy could be effective in reducing the overall propellant required for MOI.

The same technique could be applied at Earth. Instead of requiring *Adeona* to return all the way to LEO, it would be sufficient to reach the highest possible elliptical orbit that can be reached by *Newton*. A Falcon 9 can deliver 13,150 kg to LEO, whereas a Dragon capsule wet mass is only 5490 kg, which means *Newton* could rendezvous with *Adeona* at an orbit much higher than LEO. *Newton* should be able to safely descend from this orbit without the crew experiencing significant g-forces; especially compared with Mars Direct and the DRA, in which the crew vehicle enters Earth's atmosphere directly.

After the crew have descended to Earth, *Adeona*'s orbit could then be gradually reduced to LEO using aerobraking over a period of months. There are approximately 22 months between the arrival of the *Adeona* back at Earth and its next departure for Mars — ample time for this process.

Efficient pathways

Astrophysicists refer to the "Interplanetary Transport Network" (ITN), a collection of gravitationally-determined pathways through the Solar System that require minimum energy for a spacecraft to follow. Because the objects in our Solar System are constantly moving in relation to each other, these pathways are always changing. The relative positions and velocities of Earth, Mars, the Sun, the Moon, Phobos, Deimos, and perhaps even Jupiter, must be considered when calculating the optimal plan for when and where to burn propellant in order to most efficiently reach Mars and return safely back.

A related approach known as ballistic capture is a method of reaching Mars whereby the spacecraft flies to Mars orbit ahead of the planet and waits for Mars to catch up. This is estimated to be capable of reducing the amount of propellant required by as much as 25%; however, it significantly increases the duration of the trip, which is undesirable for the crew, although it could be acceptable for cargo. Also, the spacecraft is captured into a fairly high orbit, and an additional burn is still needed to reach LMO or the surface of Mars.

Predeployed assets could fly to Mars on minimum-energy Hohmann orbits, or make use of ballistic capture, as predeployed assets can take longer to get to Mars. These low-energy pathways would probably only be considered for crews in the event of breakthroughs in radiation shielding, artificial gravity, or healing/preventing the deleterious effects of microgravity.

Even when the space environment is made safer, these options may not have much appeal for crew transfer, and it's more likely that technological advancements will be applied to reduce trip times. After all, the goal is not to be in space, but on Mars (or wherever the destination may be). Mission designers will most likely seek to reduce the amount of time crews spend in space as much as possible.

Optimising the assembly orbit

Because the sections of *Adeona* to be launched by SLS weigh much less than the capacity of those vehicles (about 25 tonnes each compared with the 70 tonne capacity of SLS 70t), a more efficient approach could be to assemble *Adeona* at a higher altitude.

The trade-off would be smaller EMPPs, since, if a Falcon Heavy can launch 53 tonnes to LEO, then the amount it can launch to a higher orbit will be less (e.g. it can only deliver 21 tonnes to GTO). This could mean additional propellant launches, or it could mean fewer, since *Adeona* will require less propellant for TMI if starting from a higher orbit.

Further analysis is necessary to determine the optimal orbit for assembly and propellant loading.

Slow climb between LEO and HEEO

An SEP tug could be used to gradually pull *Adeona* from LEO to HEEO prior to TMI, and back down from HEEO to LEO after capture into Earth orbit. This could eliminate a significant fraction of the propellant — perhaps as much as half. SEP engines have very high I_{sp}, but very low thrust, which means this part of the trip would be very slow; however, the crew do not need to be aboard during this time. *Adeona* could be constructed and filled with propellant on LEO, requiring only one launch for the vehicle, and perhaps 5-6 for EMPPs.

The tug would tow *Adeona* to the higher orbit over a period of weeks or months, traversing a series of increasingly elliptical Earth orbits until the desired apogee was reached. It would then disconnect from *Adeona* and begin a steady process of descending back to LEO in preparation for the next mission.

The crew would launch in a Dragon capsule and rendezvous with *Adeona* at perigee. The maximum altitude of the HEEO will be determined by the capabilities of the Falcon 9. The Falcon 9 is capable of delivering a payload of 4850 kg to GTO, and the wet mass of a crewed Dragon with a payload of 6 crew in marssuits is about 6 tonnes (although this will vary due to the modifications necessary for Mars descent), suggesting a rendezvous orbit somewhere below GTO.

On return to Earth, *Adeona* would again capture into the highest possible HEEO that can be reached by *Newton*, and the crew would descend to Earth from there. The SEP tug would rendezvous with *Adeona* and gradually pull it back to LEO during the subsequent months.

This idea may seem appealing due to the huge savings in propellant and associated launch costs, but the expense of developing, constructing and operating the SEP tug could potentially dwarf any cost savings due to propellant reduction. Nonetheless, SEP tugs in Earth orbit could potentially serve a multitude of uses over the coming years, as humanity expands into space, and should perhaps be developed anyway.

A two-stage vehicle

In the basic design for *Adeona* presented, 6 of the 10 EMPPs are used purely to raise its orbit out of LEO. The vehicle that flies to Mars is therefore about twice as long and somewhat heavier than it needs to be. It isn't massively heavier, because the six empty EMPPs have been jettisoned, and the structure mass is kept low by using composite and/or nanostructured materials. Nonetheless, it would be worthwhile to analyse how a two-stage vehicle would compare in terms of overall cost and performance.

The first stage would likely have 5-7 EMPPs and five engines, and would serve to raise *Adeona*'s orbit from the 400 km circular assembly orbit to a HEEO with an apogee somewhere in the range of 100,000-200,000 km. At this point stage separation would occur.

The second stage would be comprised of the THAB, 3-4 EMPPs and three engines, which should be sufficient to fly to Mars and back.

For this design to be superior then there must be a reduction in propellant *and* both stages of the vehicle would have to be reusable. The first stage, having separated, could possibly lower its orbit via aerobraking over the subsequent months after stage separation, until it again reaches the assembly orbit. After inspection, it would then wait on orbit until the return of the second stage, where they would be reconnected, and the vehicle refilled with propellant in preparation for the next mission.

If the only advantage of this design over the single-stage version is perhaps 1-2 fewer EMPPs and the associated Falcon Heavy launches, saving perhaps $20 million at the most, then it's questionable whether it would be worth the additional complexity, both in the overall architecture, and in development and fabrication of the vehicle. However, a more detailed study of this idea could be fruitful.

Optimise calculations

To get maximum value from the propellant it should be burned when the spacecraft is moving fastest, as this will produce the greatest increase in the spacecraft's velocity. This is known as the Oberth effect. The higher the spacecraft's velocity during a burn, the more kinetic energy will be transferred to the spacecraft and the less to the propellant. If the spacecraft is in an elliptical orbit then the best time to burn the propellant is at periapsis, because this is when it's moving fastest. In this way, less propellant is required to achieve a given Δv.

The Δv map on which the above design is based already incorporates the Oberth effect. However, such maps are only approximate, as are the resulting calculations, and there is always room for optimisation.

Delta-v maps and the technique of "patching conics" (adding and subtracting Δv quantities in order to determine net Δv for spacecraft manoeuvres) are an approximation based on simplified two-body mechanics and well-understood Keplerian physics. This is fine for BOTE calculations, however, there are actually millions of objects in our Solar System, all affecting each other gravitationally in a complex n-body system for which the mathematics are significantly more complex.

Specialised trajectory analysis software that takes into account the Oberth effect, gravitational effects of multiple planets and moons, Lagrange Points, low energy pathways such as ITN and ballistic capture, and other factors, is necessary to identify the optimal times, places, angles and quantities for propellant combustion in order to minimise the overall requirement within the parameters of the mission.

9.9. Artificial Gravity

A trip to Mars based on a long-stay, or conjunction-class, mission profile using ordinary chemical propulsion involves approximately a year in space: 6 months outbound and approximately the same duration on the return leg. A short-stay mission requires around 20 months total in space. Either way, this is a significant amount of time to spend in a microgravity environment.

Only two people, both Russians, have spent more than 1 year in space: Valeri Polyakov, who achieved 438 days in 1994-5, and Sergei Avdeyev, who achieved 380 days in 1998-9. Therefore, only a small amount of data has been collected about such long-term exposure to microgravity, and much of it is more than a decade old. Although a variety of techniques for mitigating the adverse effects of microgravity have been developed during the past decade, minimal data is available about spending a year or more in microgravity during which astronauts had access to these measures.

In addition to a year in space, the crew spends 1.5 years on the surface of Mars, which is also a reduced gravity environment (0.38 g). Living on Mars surface will reduce the load on the body by 62% and is therefore likely to have effects proportionally similar to microgravity.

The primary concern is that the crew will spend approximately 6 months in microgravity during the outbound trip, arrive at Mars surface, and be incapable of useful work. Blue Dragon requires the crew to transfer from a landing capsule to the CAMPER on arrival, and then to SHAB, so it's even more critical that they be able-bodied on landing.

An almost equally important concern is that the crew will return to Earth after 2.5 years in reduced gravity environments and be unable to recover. Apart from the undesirable effects on beloved interplanetary explorers and heroes, such an outcome would hardly add to the glory of space exploration and may become a deterrent to future human exploration and settlement.

The problem of microgravity in an HMM is hardly a minor one. Along with the health effects of radiation in interplanetary space, it's considered by many one of the main risks associated with sending humans to Mars.

9.9.1. HMMs and Artificial Gravity

Because of these concerns, some Mars mission designers include AG (Artificial Gravity) in their architecture. The simplest way to create a gravity-like effect is with centrifugal force, which can be produced by spinning part or all of the spacecraft. Unfortunately centrifugal force is not a perfect substitute for actual gravity, and it's not yet known for certain if it will be an adequate substitute.

The DRA does not specify unequivocally whether AG should be used, although it does state that AG would be far more important for a short-stay, opposition-class mission due to the longer time spent in space. It is not considered essential for a conjunction-class mission, and the MTV designs in the DRA do not include support for AG.

From the DRA:

> Adverse physiological changes due to reduced gravity may be prevented by exposure to some level of artificial gravity, but the specific level of gravity and the minimum effective duration of the exposure that is necessary to prevent deconditioning are not yet known. Although artificial gravity should reduce or eliminate the worst deconditioning effects of living in zero gravity, rotating environments frequently cause undesirable side effects, including disorientation, nausea, fatigue, and disturbances in mood and sleep patterns. If artificial gravity is to be employed, significant research must be done to determine appropriate rotation rates and durations for any artificial gravity countermeasures. The decision on whether artificial gravity must be employed to adequately support crews on their transits to and from Mars, as well as the decision on the necessary gravity level and rotation rate, has significant implications for vehicle design and operations.

AG is therefore no panacea, and carries its own set of health-related problems.

Research indicates that the negative effects of centrifugal force are largely unnoticeable if the rotation rate is maintained below a maximum of 2 rpm. To produce Earth gravity at this spin rate requires a radius of approximately 224 metres; to produce Mars gravity requires a radius of about 85 metres.

The use of AG in Blue Dragon would significantly increase the cost of the mission due to additional mass of both spacecraft and propellant, and increased complexity of propulsion, navigation, communications and power systems. It introduces additional hazards and failure modes, and would significantly increase development costs. It would also compromise the interior design, as a spaceship designed for microgravity can be made more space efficient.

Complexity, mass and cost need to be minimised, especially for early-stage HMMs. The question is whether the considerable cost associated with AG would produce sufficient ROI compared with a regimen of PT (Physical Training), nutrition and medication.

9.9.2. AG in Mars Direct

From Mars Direct:

> Artificial gravity is provided to the crew on the way out to Mars by tethering off the burnt out Ares upper stage and spinning up at 1 rpm.

To create Mars gravity at a spin rate of 1 rpm requires a radius of rotation of about 338 m. At this spin rate the adverse side-effects of centrifugal force should be negligible, which is good. However, consider what this idea means for the architecture:

- The habitat, which, in Mars Direct, is used for the outbound trip as well as surface habitation on Mars, is designed for a gravity environment: AG in space and Mars-g on Mars. It will therefore have a floor, ceiling and walls, with cupboards, screens, controls, etc. mainly on the walls. The result is a less compact and heavier spacecraft, or one with fewer fixtures/features.

- Although designed for a gravity environment, after launch the hab will be in microgravity until the AG is set up, which is not until after TMI, and must also return to microgravity mode after being spun down prior to MOI. In addition, if the AG system fails and the tether must be dropped for any reason, the hab must be able to continue to Mars in a microgravity mode. Designing the hab to be operable in both AG and microgravity modes will add complexity.

- Because the tether is not rigid there's a risk the counterweight could crash into the hab, and it may not be possible to prevent this even by dropping the tether.

- The mass of the 1500-metre tether must also be launched, which means more mass, more propellant and more cost. Note that the tether must be strong enough to safely handle the tension produced by centrifugal force (equivalent to Mars gravity times the mass of the hab), plus a suitable safety margin.

- Communications with Earth are more challenging with a spinning spacecraft, as it's more difficult to keep antennas pointed in the right direction without sacrificing signal strength. Communications with Mars are already up to one million times more challenging than with the Moon, since Mars is up to 1000 times as far away (signal strength decreases with the square of the distance).

- With a spinning spacecraft it's more difficult for navigation sensors to track the position of Earth, the Moon and stars.

- Collecting solar energy with solar panels is more difficult and possibly less efficient, as they should ideally be always directly facing the Sun.

- The hab and the counterweight both require an RCS, which must work in tandem in order to produce a stable spin around a common centre of gravity, or for any course manoeuvres. This would require advanced avionics.

- RCSes, comprised of thrusters and propellant (i.e. mass), are required on both the hab and the counterweight, in order to spin the whole assembly up and then down again prior to MOI.

- Additional propellant is required to send, not just the hab, but also the counterweight, tether and associated systems on TMI.

- Course corrections and other manoeuvres are more difficult, as both the spacecraft and counterweight must be pushed in the same direction at the same time, without overstressing and breaking the tether, or causing it to stretch and rebound, or causing it to become slack. Plus, the dynamics of the spinning masses must also be accounted for.

- As the assembly approaches Mars it must be spun down and the tether dropped before MOI. This will put the spacecraft on a trajectory that is difficult to predict in advance with precision, making it more difficult to calculate, in advance, the required burn to place the spacecraft on the desired EDL trajectory. It may have to be re-calculated in real time, which is riskier. Yet the hab must hit the atmosphere at exactly the right angle and altitude. EDL is already a very technically challenging and dangerous process, and this additional factor would make it even more difficult.

- If the tether breaks for some reason (e.g. meteoroid impact, overstressed) *en route*, the hab and counterweight will fly off in opposite directions with a velocity component perpendicular to the desired trajectory. The hab would need additional propellant to make a course correction if this happens. In addition, MCC must be prepared to track and communicate with the hab (which may be on an unknown trajectory) and assist with on-the-fly calculations to determine course correction manoeuvres.

- Only Mars-level gravity is provided, which is still a reduced-gravity environment, and the crew will still experience adverse health effects to some degree. Also, the AG solution described in Mars Direct is only for the outbound trip. The crew may arrive at Mars adapted to Mars gravity, but on return to Earth will be in almost the same condition as if no AG was provided at all. Therefore, despite considerable additional expense and complexity, this strategy only goes about 19% (38% * 50%) towards addressing the problems of low gravity associated with a human mission to Mars.

Some of these issues can be addressed by using a rigid telescopic truss instead of a tether, but that would have a high mass and therefore be an even more expensive solution (see _Athena - a 12-person MTV with AG_).

9.9.3. Better Living Through Chemistry (and Training)

The benefits of AG don't warrant the cost and complexity of implementation, particularly for the first few missions, and it isn't a necessity. Many astronauts have spent months in microgravity; on return to Earth, some are able to walk away from the spacecraft and some are not, but all recover and readapt to Earth gravity in time.

The crew of an IMRS mission will be able to offset the deconditioning effects of microgravity by spending some time each sol doing exercise, either in *Adeona*'s gym or by going on EVA. Their diet will be high in protein and minerals.

The astronauts may also be administered medication to mitigate bone loss, as successfully demonstrated by NASA and JAXA in 2011 (LeBlanc et al. 2014). Through administration of bisphosphonates, a class of drugs commonly prescribed to prevent bone loss and treat osteoporosis, bone density loss in the femur was only 1%, and hip bone density actually increased by 3%. Without, the average bone density loss was 7% in the femur and 5% in the hip. The research concluded:

> *These results indicate that the use of an antiresorptive plus exercise may be effective during long-duration spaceflight and that bone measures at 1 year remain at or near baseline values.*

By incorporating modern physical training principles and equipment, and appropriate nutrition and pharmaceuticals, the crew should be fully able, even after spending 6 months in microgravity.

On arrival at Mars it's understood that the crew will need at least a few sols, perhaps even a week or so, to adapt to Mars gravity. This is a clear advantage of a long-stay architecture. As the crew has about 18 months on the surface, this comparatively brief recovery period will significantly not impact the surface mission. The astronauts will naturally be eager for EVA, and will push themselves to recover quickly. In fact, going EVA could be the most effective way for them to regain strength.

Another reason why microgravity may be preferred to AG is simply that it's more fun, which is good for morale, and for entertaining people back on Earth. Microgravity is arguably half the fun of being in space, after all.

The Inspiration Mars mission to send two people on a Mars flyby will require them to spend more than 1.5 years in microgravity. If successful, this mission will go a long way towards answering the question of whether it will safe to send a crew to Mars surface and back on a 2.5 year mission. Mars One also does not make use of AG.

9.9.4. Health and Fitness

While not a complete antidote to the effects of microgravity, the most important strategy will be exercising every sol.

Resistance training

The human body responds to loading. When it becomes unloaded — for example, by spending time in a reduced gravity environment — the body adapts by losing muscle and bone mass, simply because not as much is needed. When it is loaded — for example, by lifting weights several times per week — the body adapts by increasing muscle and bone mass.

Although there's a vast body of knowledge around fitness and bodybuilding, once you trim away all the cruft there are really only a few basic principles:

1. The body responds to loading. This is a function of two things:

    •   The amount of weight it's loaded with. The more weight the body is loaded with, the more it will tend to grow stronger.

    •   The amount of time it's loaded (also called "time under tension"). The more time the body spends loaded, the more it will be forced to adapt.

2. Nutrition is a critical factor, especially protein. Protein should represent a significant fraction of the diet, as it protects against muscle catabolism and provides the building blocks (amino acids) to synthesise new muscle tissue for growth and repair. A balance of other nutrients is also essential: healthy fats, complex carbohydrates, vitamins, minerals and plenty of water.

Strength training involves performing a range of exercises to train all the muscles in the body. But is there an exercise that simulates loading due to gravity? Yes, and this exercise is well-known to bodybuilders as being the number one exercise for loading almost the entire body: squats.

Squats are known as the king of exercises because they build the full kinetic chain of the body. They're a highly functional exercise, requiring the muscles used in standing, walking, jumping and running. Squats load the body in almost the same way as normal gravity — vertically downwards — but to a greater degree. Squats may yet be revealed as the number one exercise for reversing the deconditioning effects of microgravity.

The intention for Blue Dragon is that the crew spends approximately 1 hour per sol doing strength training. Not only squats, because that would be a bit boring and would lead to over-training, but all the so-called "Big 5" exercises: squats, deadlifts, chest press, shoulder press and chin-ups.

Spending only 4% of each sol with the body partially loaded may not seem like enough to offset the effects of microgravity. However, muscles do not need to be overloaded for long periods to stimulate growth, and many people achieve muscle growth with just 2-3 gym sessions per week. Regular, short periods of intense exercise are effective at improving strength and fitness.

The intent behind resistance or strength training is usually muscle growth, fat loss or both. However, one of the major problems with microgravity is loss of bone density, and fortunately strength training helps with this, too. Research shows that resistance training produces noticeable increases in bone density, and is therefore sometimes recommended for osteoporosis sufferers. Other conditions that improve as a result of strength training include cardiovascular disease, diabetes and high blood pressure.

Rather than lifting weights, which would be ineffective in microgravity, *Adeona* will be equipped with multi-functional hydraulic gym equipment. Plenty of commercial gym equipment based on hydraulics is available, which will work perfectly well in a microgravity environment.

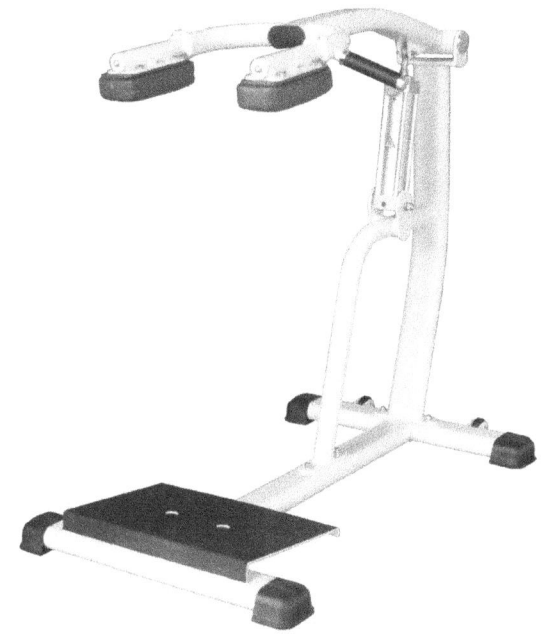

Hydraulic squat machine (Credit: Qingdao Passion Fitness Equipment Co.)

Yoga

Something else that astronauts may be able to do in microgravity to stay fit is yoga, which produces a wide range of benefits for both body and mind. Although many yoga postures require gravity, particularly those related to balance, others could be adapted for a microgravity environment. Improved flexibility and breathing, cleansing the lymphatic system, releasing stress stored in the spine and muscles, and all the other benefits of yoga could potentially be gained by a Mars mission crew.

Cardiovascular exercise

Because it's easier to push blood around the body without gravity, the heart doesn't have to work as hard, which causes a decrease in heart mass. This can be mitigated by regular cardiovascular activity. The heart must be exercised like any other muscle, but it's exercised differently, using cardiovascular exercise instead of resistance training. The heart rate must be elevated. Although this is achieved for intermittent periods with strength training, cardiovascular exercise serves to sustain an elevated heart rate for longer periods.

EVA on Mars

While on Mars for 1.5 years, the crew will experience considerable additional exercise in the form of regular EVAs; up to about 32 hours per week. During this time they'll be wearing a marssuit with a mass of approximately 20 kg.

While this is not a heavy load, particularly in Mars gravity (where it will feel like just 7.6 kg), the astronauts will be performing a range of physical activities whilst wearing the suit, including hiking, bending, kneeling, standing, lifting and carrying. Although the gravity on Mars is reduced, inertia is unchanged, and muscles will be constantly activated while in the suit in order to maintain balance.

Going on EVA could therefore be an important part of the necessary physical conditioning to mitigate bone and muscle atrophy. The astronauts will surely prefer going on EVA on Mars to doing squats in the hab.

IMRS program

The health and fitness program for IMRS crews may look something like this:

- In the THAB:

    - 3 sols per week: cardio.

    - 3 sols per week: yoga in the morning, strength exercises in the afternoon.

- In the SHAB:

    - 4 sols per week: yoga in the morning, then EVA.

    - 2 sols per week: cardio in the morning, strength exercises in the afternoon.

- 1 sol per week: rest.

- 30 minutes breathing meditation every day.

- Plenty of protein to minimise muscle loss and support muscle growth and repair.

- Plenty of minerals, especially calcium, magnesium and iron.

- Bisphosphonates to mitigate bone loss.

- Plenty of sleep and clean water.

9.9.5. *Athena* - a 12-person MTV with AG

Assuming that the IMRS program is continued beyond its first decade, developments in technology could make it affordable to construct a larger ship that provides Mars-level AG on the way to and from Mars. The following design may be effective:

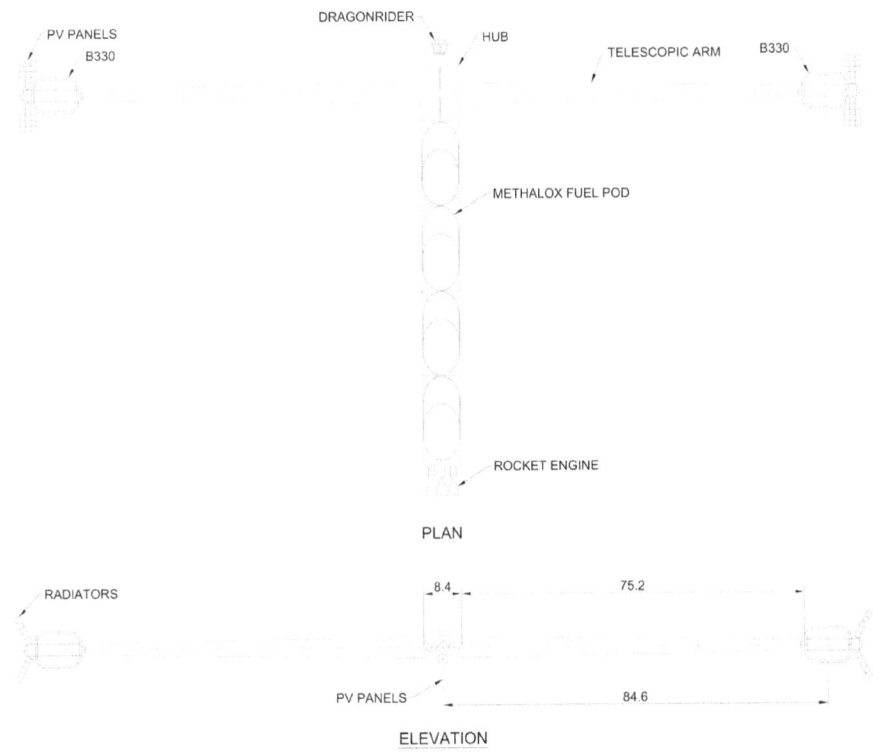

Athena - concept for a 12-person Mars Transfer Vehicle with artificial gravity

This MTV is named "Athena", after the Greek goddess of wisdom, courage, inspiration, civilisation, law and justice, mathematics, strength, strategy, the arts, crafts, and skill. Athena is the companion of heroes and patron goddess of heroic endeavour.

This design has two THABs based on B330 modules, and would support a crew of 12. At least two SHAB's would therefore be needed at the IMRS to support crews of this size. Although this vehicle would be considerably more expensive than *Adeona*, expanding the missions to support larger crews may make them cheaper per astronaut, and the addition of AG will improve astronaut health.

The central core of the vehicle, comprised of a central hub plus four large EMPPs (sized for the payload of the SLS or other SHLLV) would be built first. The two telescopic arms are designed to fit inside the payload fairing of the Falcon Heavy when collapsed. They would be launched by Falcon Heavy, connected to the central hub, and the B330 modules attached to their ends. The assembly would gradually be spun up to 2 rpm using the RCSes on the B330 modules, causing the arms to extend under the influence of centrifugal force.

The joins between the sections of the arms would be bolted or clamped together (ideally automatically, but perhaps manually by astronauts on EVA) to form airtight tunnels that would permit crew members to move between the THABs and the central hub.

The radius of rotation to the centre of the B330 modules is 84.6 metres, which, at 2 rpm, will provide Mars-level gravity.

The front of the hub would provide a docking port for a Crew Dragon capsule. In order for a capsule to dock, the docking port would be able to counter-rotate in order to remain stationary relative to an approaching capsule. After docking, the connected capsule and docking port would gradually be spun up to 2 rpm to match the spacecraft, so the crew could enter the hub, then transfer through the arms to the THABs. Inside the hub would be a microgravity environment, with the apparent gravity level increasing with distance from the hub.

Once the spacecraft is spun up, it would not need to be spun down again, because crews can enter and exit via the central hub. GNC and communication systems would be attached to the central hub where they will function more effectively, being easier to point at the stars or Earth respectively.

If astronauts are able to tolerate higher rotation rates, or if the arms can be made longer, then the outbound journey could begin with Earth-level gravity and decreased to Mars-level gravity during the outbound journey by progressively reducing the spin rate. This would permit a gradual and stress-free adaptation to Mars gravity. On the return journey the process would be reversed, and the astronauts should, in theory, arrive back at Earth more-or-less fully recovered.

10. Mars Ascent Vehicle

The MAV will be a VTOL vehicle capable of pinpoint landing on Mars, refilling with propellant from local Martian resources, and returning the crew of six astronauts to Mars orbit. Instead of bringing H_2 or CH_4 from Earth, the MAV will use H_2O obtained from the Martian surface in addition to CO_2 from the atmosphere to manufacture 100% of the required ascent propellant.

This strategy should lighten the landed vehicle to the extent that a common stage can be used for both descent and ascent, thus further reducing the MAV's mass, making it far cheaper and easier to deliver to Mars. Nanostructured materials, composite cryotanks and other innovations are incorporated into the design to reduce the vehicle's mass as much as possible, to the point where the necessary power for manufacturing ascent propellant can be provided by solar energy rather than nuclear, thus reducing complexity as well as negative environmental effects.

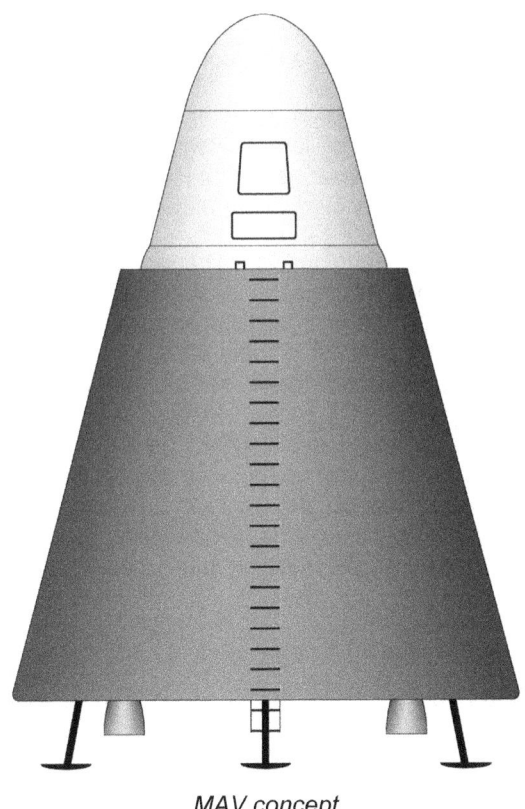

MAV concept

10.1. MAV Concept

The MAV concept presented in this section is somewhat different than others that have been presented in the past, being lighter, simpler, more technologically ambitious, and possibly cheaper.

10.1.1. Elements of the MAV

Drawings included in this section are simple and intended only to help illustrate some basic design concepts for the MAV. Complexities of the ISPP subsystems, propellant pumps, thermal control systems, avionics, etc. are not shown.

Kepler

Also known as the "Mars Ascent Capsule", the modified Dragon capsule at the apex of the vehicle is empty during descent, but during ascent to Mars orbit will contain six crew plus their marssuits and areological samples.

Because it doesn't need to land or re-enter an atmosphere, *Kepler* will be fabricated without a heat shield, landing legs, or SuperDraco engines and associated tankage. It will, however, retain its RCS/OMS and the smaller Draco engines, as these are necessary for docking with *Adeona* after ascent.

The trunk section normally connected to a Dragon capsule could potentially be included in the MAV and used to contain solar panels, which would help to supply power to the MAV for ISPP, in addition to powering the Dragon's systems while in Mars orbit. Another use for it could be to carry samples. However, the trunk would add significant ascent mass, and is best eliminated. Solar energy can be provided by the MAV's PV coating and/or PV blanket, and since *Kepler* only needs to carry six astronauts instead of the usual seven for which it is designed, the lower centre seat can be replaced with a storage area for samples. This may mean fewer samples, but the total sample mass is limited anyway due to mass constraints of the payload; plus, the samples can be returned to Earth with the crew at the end of the mission more easily if *Newton* is similarly configured.

Superstructure

The lower section of the MAV is currently envisaged to be a frustum of a cone, and manufactured from a high-performance aerospace material such as an aluminium-lithium or titanium alloy, or titanium-carbon composite. Compared with the vertical cylinder design typical of rockets, the conic shape lowers the centre of gravity of the vehicle and provides a wider base, which will increase stability should the MAV land on uneven terrain.

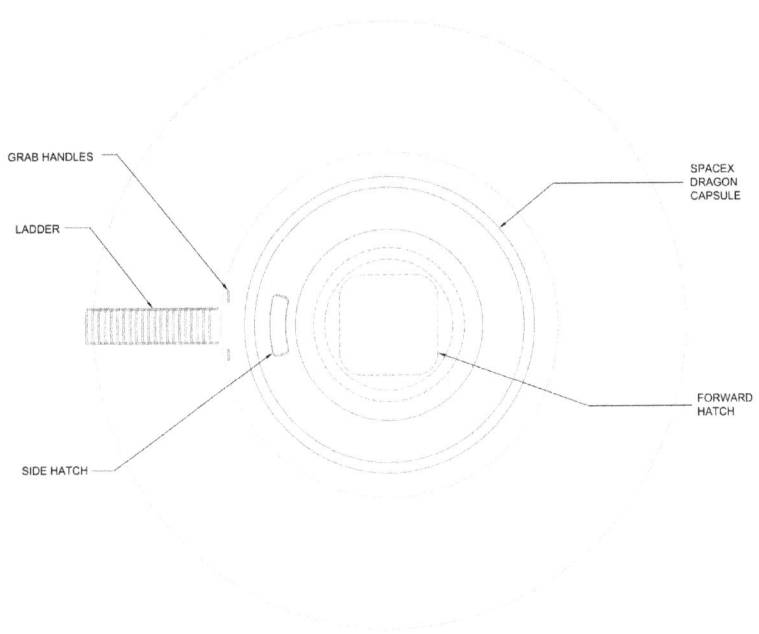

GRAB HANDLES

LADDER

SPACEX
DRAGON
CAPSULE

FORWARD
HATCH

SIDE HATCH

PLAN VIEW

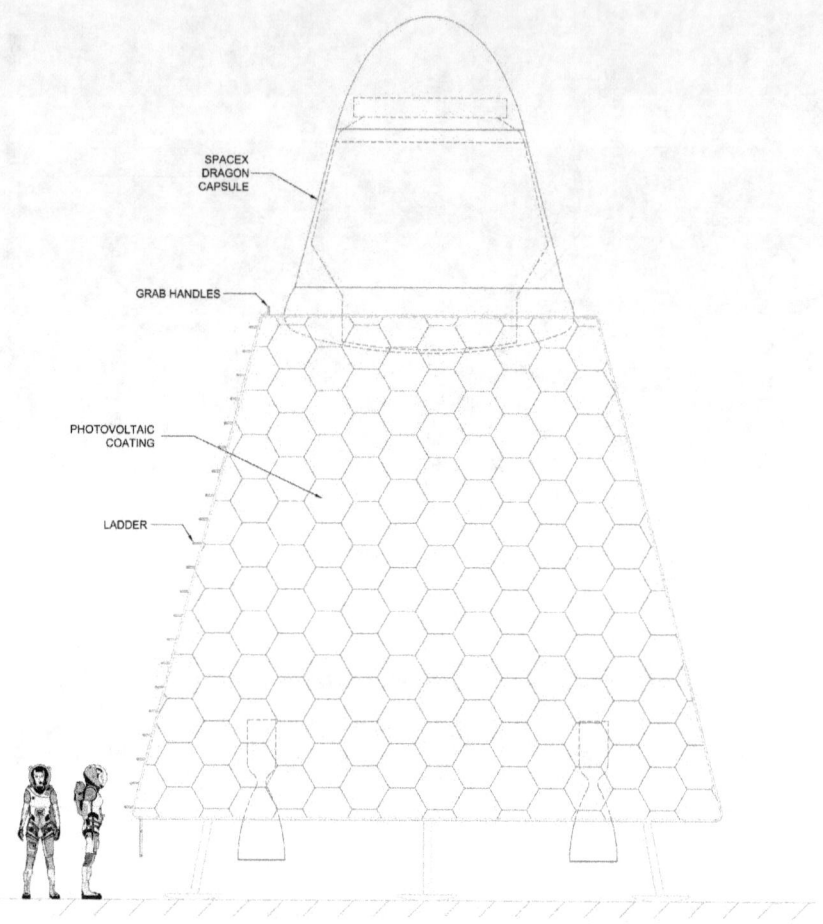

SPACEX
DRAGON
CAPSULE

GRAB HANDLES

PHOTOVOLTAIC
COATING

LADDER

SIDE ELEVATION

Photovoltaics

One of the reasons for the ERV's nuclear reactor in the Mars Direct architecture is that propellant production can proceed quickly, day and night, in order to minimise hydrogen boil-off and ensure that all the propellant gets made. However, by using local water as a hydrogen source, the hydrogen immediately converted to methane and does not need to be stored. Thus, there will be no boil-off and less urgency to manufacture the propellant as quickly as possible.

A constant power supply is therefore not a strict requirement for ISPP. Solar energy could be used, and propellant manufactured only during the daytime when power is available.

There are about 20 months between when the MAV arrives at Mars and when the crew leaves Earth at the subsequent launch opportunity. Even if propellant is only being made during the day, this should still be plenty of time to ensure that the MAV is full of propellant before the crew leave Earth. In fact, there is about 44 months between the MAV's arrival and departure (26 between its arrival and the arrival of the crew, plus 18 months of surface mission), so extra time is available if required.

By using solar panels instead of a nuclear fission reactor for ISPP power, the mass of the reactor, robotic truck and electrical cable that would connect the reactor to the MAV, are eliminated, as is any risk of an unshielded reactor irradiating the MAV, the crew, and the pristine and scientifically-valuable Martian environment.

One approach for providing power for ISPP could be via a roll of solar fabric (the "PV blanket") stored in the lower section of the MAV, on a fixed roller, which the AWESOM robot can unroll on arrival simply by dragging one end away from the MAV. Another method for supplementing MAV power is for the outer surface to be coated in organic photovoltaic paint, such as that being developed by NanoFlex Power Corporation and the University of Newcastle. Yet another idea would be for the outer walls of the MAV to open up like a flower, revealing solar panels on their inner surfaces.

In the event that photovoltaics cannot produce sufficient power to manufacture the necessary propellant in the given time frame, a possible supplementary power source would be an ASRG. This would generate heat in addition to electricity, which could help maintain the Sabatier reactor at its required operating temperature of around 300-400 °C, and to keep the electronics warm. In fact, an ASRG may even be preferable to the photovoltaic blanket, which would cover useful ice-containing ground, and could interfere with operation of the AWESOM robot. If radioactivity is a concern, the ASRG can be removed along with the ISPP modules before launch of the MAV, and buried using the CAMPER's excavator attachment.

Descent/ascent engines

Several types of methalox rocket engines have been proposed and developed, however, none are currently in active use.

One possible candidate for MAV engines could be the XR-5M15 developed by XCOR, which provides about 33.4 kN of thrust and weighs about 48 kg. At least six would be required for Mars ascent, giving a total thrust of about 200 kN and a total mass of about 286 kg. However, even with six of these engines the acceleration would be rather low and the engines would need to burn for a long period to reach orbit. More could be used, but four engines work better for the geometry of the MAV, and larger engines typically provide a superior TWR.

The only other methalox engines in development are either too small or too large.

Therefore, although it will incur some development costs, it will probably be necessary to develop another new methalox engine.

The engines, which will be used for both descent and ascent, are estimated to have a thrust of approximately 90 kN and I_{sp} of approximately 380 s. These figures are similar to the engines described in the DRA, which have an average thrust of approximately 22,000 lbf (88,964 N) and I_{sp} of approximately 369 s. The higher I_{sp} is likely to be achievable for a newly-developed methalox engine incorporating the latest materials and manufacturing techniques.

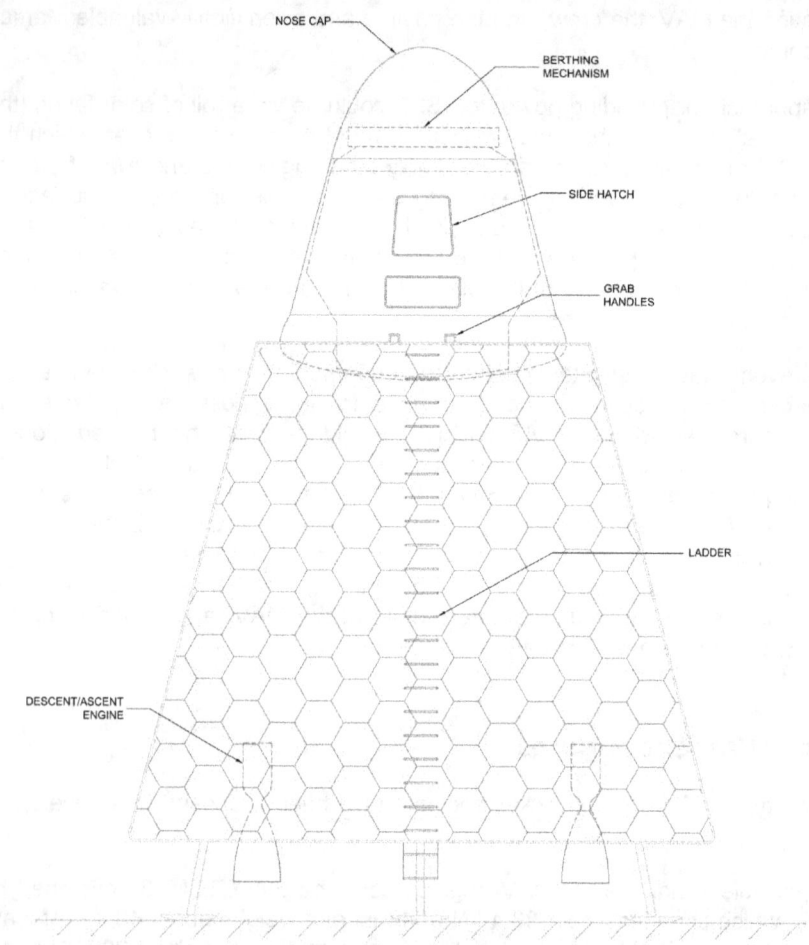

FRONT ELEVATION

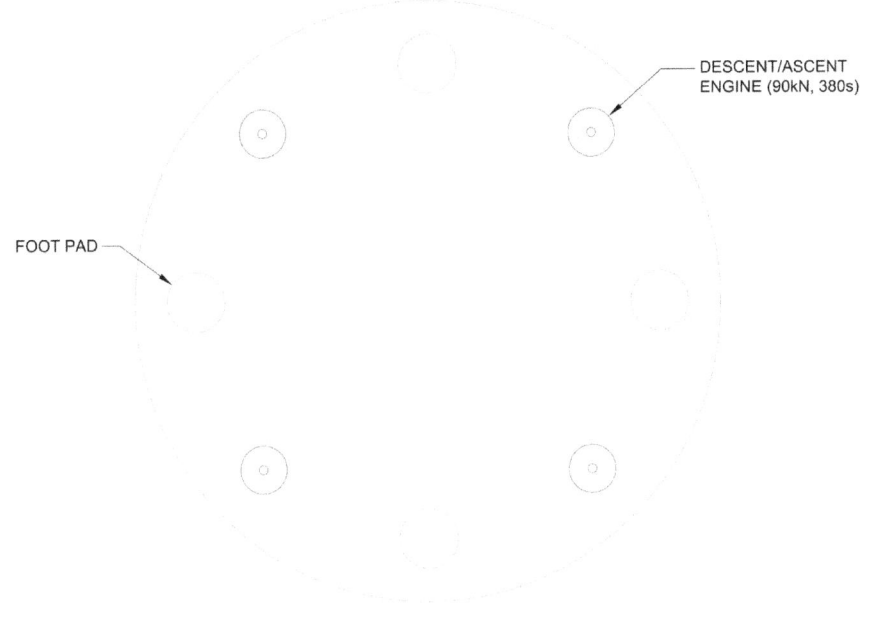

DESCENT/ASCENT
ENGINE (90kN, 380s)

FOOT PAD

UNDERNEATH VIEW

ISPP system

The lower section of the MAV includes equipment for manufacturing methalox bipropellant from resources available on Mars. This includes:

- A pump to collect atmosphere.

- A dust filter.

- A bed of zeolite 3A to absorb water from the atmosphere.

- A condenser to separate CO_2 from the dried air.

- A roll of photovoltaic material to provide power.

- The AWESOM robot to collect water from the surrounding terrain.

- A water tank.

- An electrolysis unit to separate water into H_2 and O_2.

- A Sabatier reaction chamber to react H_2 and CO_2 to produce CH_4 and H_2O.

- A heater to provide heat to the Sabatier reaction chamber, water tank and electronics.

- Additional elements as necessary for separation of gases, etc.

These components and their operation are discussed further in the section on <u>In Situ Propellant Production</u>.

In the drawings below, the modules are shown as neat rectangular units. This is really just concept art. The individual components of the ISPP system will be quite varied and complex in their actual designs.

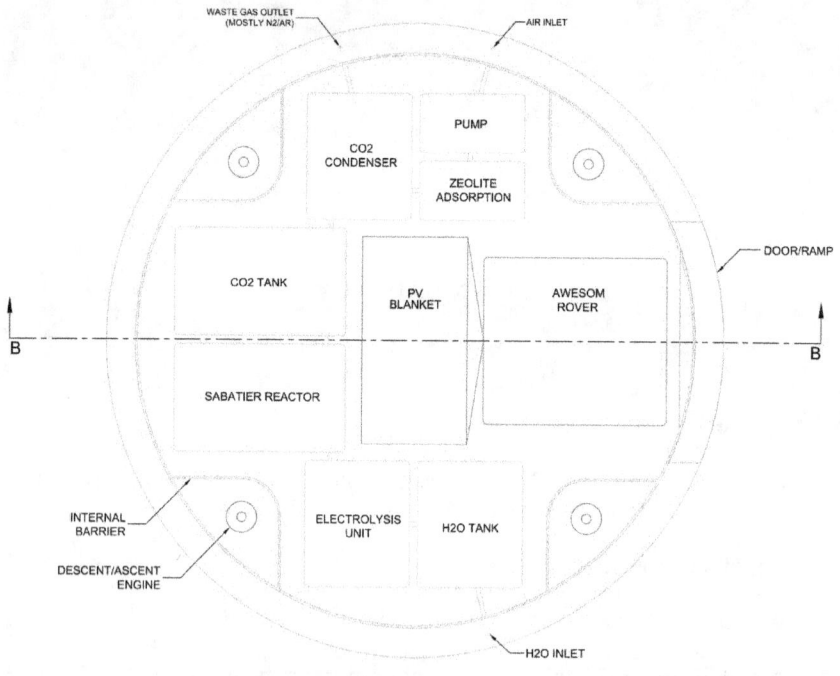

SECTION A-A

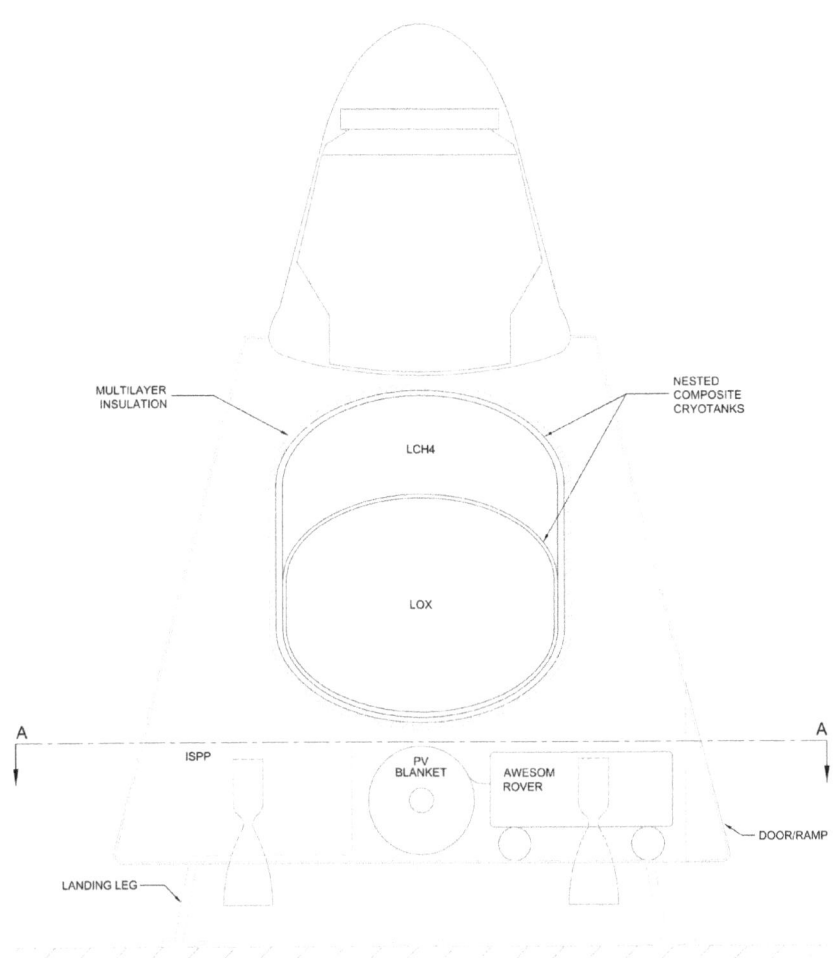

MULTILAYER
INSULATION

NESTED
COMPOSITE
CRYOTANKS

LCH4

LOX

A   A

ISPP

PV
BLANKET

AWESOM
ROVER

DOOR/RAMP

LANDING LEG

SECTION B-B

Propellant tanks

The same basic design concept of nested composite cryotanks as used in *Adeona* will also work for the MAV.

Avionics and imaging

The MAV avionics include a sophisticated GNC system based on LION or a similar technology, which will be capable of pinpoint landing on Mars by visual identification of landmarks. The MAV will have cameras in its base, sides and

203

attached to the capsule.

10.1.2. Common Descent-Ascent Stage

With regards to landing masses of 20+ tonnes on Mars, the DRA concluded the following:

> The reference EDL architecture that was ultimately selected for this study was a hypersonic aeroassist entry system, with a mid lift-to-drag ratio (L/D) aeroshell that was ejected at low supersonic Mach numbers. An LOX/LCH4-fuelled propulsion system was used for deorbit delta-V maneuvers, RCS control during the entry phase, and final terminal descent to the surface.

Thus, the DRA proposes aerobraking plus methalox propulsion for landing large items on Mars. This approach is retained in Blue Dragon.

Because ascent propellant is being made on Mars, and both descent and ascent propulsion systems are based on methalox, it should be possible to use a single stage for both ascent and descent. If so, this will reduce the complexity and mass of the overall vehicle, making it simpler to land on Mars while not significantly increasing ascent mass.

The usual reason for having separate stages for descent and ascent is to improve the efficiency of ascent. A spacecraft is typically heavier during descent because it carries at least some of the ascent propellant, plus its payload. By leaving behind the payload plus any superfluous engines, tankage and superstructure required only for descent, the ascent mass can be minimised and thus far less propellant is required for ascent (which means less mass needs to be landed on Mars, which is easier and cheaper).

For example, the descent stage of the Apollo Lunar Excursion Module included engines and propellant tanks in addition to equipment such as tools and experiments to support surface activities — essentially everything that was not needed for ascent. Leaving this hardware behind enabled the ascent mass to be much lighter.

In Mars Direct, the ERV is landed with 6 tonnes of H_2. In the DRA, the MAV is landed with about 6.5 tonnes of CH_4 or 2 tonnes of H_2. Delivering this mass to Mars has a significant impact on EDL system design, and the initial size of the vehicle. With these architectures it makes sense to have a separate descent and ascent stage, because the descent vehicle needs to be much larger and more powerful than the ascent vehicle needs to be.

In Blue Dragon, however, the MAV lands with empty tanks, because all the ascent propellant is manufactured on Mars (see In Situ Propellant Production). The MAV will also not be carrying any experiments or items of equipment for use by the crew, as these will be delivered in the SHAB and cargo capsules. *Kepler* will be empty, so the only "payload" will be the ISPP and ISWP systems, i.e. the hardware for production of energy, water and propellant, which should mass about a tonne or two at most.

The descent vehicle therefore does not really need to be significantly bigger or more powerful than the ascent vehicle. This reduces any efficiency gains that could be obtained by having separate descent and ascent stages, to the point where it could be more efficient to simply have a single stage for both descent and ascent. The MAV would thus descend and ascend using the same propellant tanks and engines.

In order to reduce the ascent mass and thus the quantity of propellant that needs to be manufactured, the ISPP subsystems can all be removed prior to launch. The PV blanket is pulled out by the AWESOM robot on arrival, so these components are already removed. Even if a small nuclear reactor is used for MAV power as described in Mars Direct, this, too, will have already been removed from the MAV, having been relocated via a small robotic truck.

The ISPP hardware can be designed as modular units that the crew can easily unbolt from the MAV and leave behind on the surface of Mars. They may even wish to do this on arrival if the MAV's propellant tanks are already full at that point. The ISPP modules, AWESOM robot and PV blanket could be moved to the SHAB and used to make propellant for the CAMPER or a generator, and to provide the SHAB with additional energy.

With these items removed from the MAV, the only surplus ascent weight is a slight excess of superstructure, i.e. the "container" in the lower part of the vehicle where those items were originally located. The mass of this lower section can be reduced via removable panels, which would provide access to the internal modules so they can be extracted.

A single descent-ascent stage will make the whole vehicle simpler, lighter, cheaper, easier to launch from Earth, and, most importantly, much easier to land on Mars. The landed mass of the MAV in the DRA is comprised of the descent and ascent stages, all ISPP hardware, and the methane component of the ascent propellant, for a total landed mass of at least 20-30 tonnes. However, the MAV in the IMRS plan, with its common descent-ascent stage and zero propellant, lands weighing only around 10 tonnes.

Because the same propellant tanks are used for both descent and ascent, for optimal efficiency the propellant requirement for descent will be less than or equal to that required for ascent. This should be possible, because with this design the descent and ascent masses are very similar. Although the ISPP and ISWP hardware are included in the descent mass but not the ascent mass, the crew plus their marssuits, kit bags and areological samples, are part of the ascent mass but not the descent mass. In other words, the descent and ascent payloads balance out.

The MAV approaches Mars via direct entry, which will requires a Δv of about 6.5 km/s to land. This is somewhat higher than the 5.6 km/s Δv required for ascent, since the MAV must only reach Mars orbit on ascent. However, the descent propellant requirement is reduced through the use of aerobraking.

10.2. Entry, Descent and Landing

The following frames extracted from the legendary documentary "The Mars Underground" help to illustrate a possible EDL system for the MAV (although the MAV in the images is somewhat different from the proposed concept).

In the movie, the MAV uses three large parachutes to help slow its descent. However, in the DRA, as mentioned, the MAV descent is entirely propulsive, which is probably a good idea. Parachutes are extremely difficult to test, they add considerable complexity to the EDL system, and can only do so much in the thin Martian air. They could also cause the MAV to be blown off course by wind, which would interfere with the goal of accurately landing the MAV at a specific location.

The MAV approaches Mars directly from the inbound MTO, and is therefore travelling at high velocity when it hits the atmosphere. Aerobraking is used to slow its descent, converting velocity into heat via friction. At this point its propellant tanks are full.

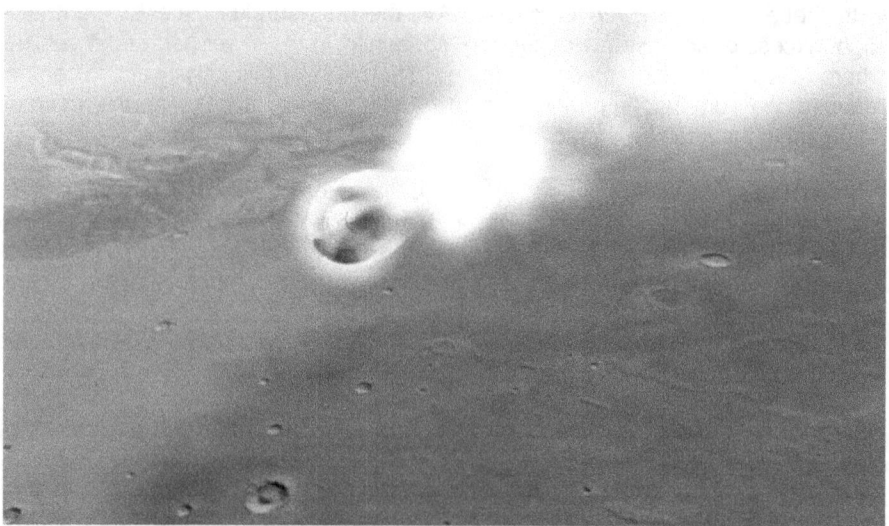

The MAV is protected by its aeroshell, which also helps to slow it down. Inflatable aeroshells, much lighter than the metal one shown in this image, are actively being developed by NASA. The most recent was tested in an experiment referred to as "IRVE-3", or "Inflatable Reentry Vehicle Experiment".

The process of aerobraking slows the MAV until its rate of descent is low enough that it can land propulsively with the available propellant.

When the MAV's speed has reduced sufficiently, the aeroshell is jettisoned.

Sensors and cameras on the underside of the MAV, in combination with signals from MCOS, enable the onboard software to coordinate an accurate landing. The final stage of EDL is fully propulsive, using the four methalox descent/ascent engines and consuming virtually all of the propellant.

10.3. In Situ Propellant Production

ISPP refers to the manufacture of methalox propellant from indigenous Martian resources. It is arguably one of the most important applications of ISRU ever proposed, because it enables astronauts to return to Earth from Mars using present-day technology.

Launching from the surface of Mars has been cited as one of the biggest obstacles when it comes to HMMs. It is for this reason that the plan for Mars One is to send people to Mars one-way, thus side-stepping this difficulty.

Launching from Mars is difficult is because of the large mass required. An estimated minimum wet mass (vehicle plus propellant) of 30-40 tonnes would be necessary to lift a crew of six humans from the surface of Mars to Mars orbit. Yet the heaviest object soft-landed on Mars to date is the *Curiosity* rover, which weighed just 900 kg.

Using the new SHLLVs such as the SLS it should be possible to land payloads of up 20-30 tonnes on Mars. However, this may still not be enough for a MAV filled with propellant. To land larger payloads on Mars will require space vehicle technology yet to be developed.

However, if the ascent propellant can be made on Mars, a MAV can be landed on Mars empty of propellant. This makes the landed mass significantly less, which means it can be done sooner. This is why the idea of ISPP was developed.

10.3.1. Background

ISPP in Mars Direct

When Mars Direct was developed by Dr Robert Zubrin and David Baker at Martin Marietta during the 1990s they described the benefits of landing an ERV on Mars carrying only a fraction of its ascent propellant, and obtaining the remainder from resources available in the atmosphere.

In Mars Direct, about 6 tonnes of H_2 is brought from Earth. Hydrogen is the most difficult element in methalox propellant to obtain from Mars, being unavailable from the atmosphere in quantity. The H_2 is reacted with CO_2 extracted from the Martian atmosphere via the Sabatier reaction, producing CH_4 and O_2, which are liquified and stored cryogenically. Additional O_2 is obtained from CO_2 via RWGS (Reverse Water Gas Shift), which produces H_2O that is then electrolysed to H_2 and O_2. A total of 108 tonnes of methalox are produced: 24 of CH_4 plus 84 of O_2. 96 tonnes are for Mars ascent, and 12 tonnes are for surface vehicles.

The hydrogen transported to Mars represents just one eighteenth of the total propellant mass required (plus a margin to compensate for boil-off).

Mars Direct showed that if the propellant for Mars ascent is manufactured at Mars in this way, the resulting mass saving could potentially reduce the cost of a human mission to Mars as much as a factor of eight. This single innovation brought the goal of a HMM from future-fantasy into near-term reality. Mars Direct gathered so much attention that the two engineers travelled the US presenting their plan and building renewed interest in sending humans to Mars.

ISPP in Mars Semi-Direct and the DRA

With feedback from NASA, Dr Zubrin modified the Mars Direct plan to create Mars Semi-Direct. This variation on the architecture used a Mars Ascent Vehicle instead of an Earth Return Vehicle, the difference being that the MAV only needs to reach Mars orbit, rather than travel all the way back to Earth. This significantly reduces the amount of ascent propellant required. Instead of 108 tonnes of methalox, only about 20-30 tonnes is needed. This reduces risk, propellant tank size and mass, energy requirements for the MAV, mass of the power system, the landed mass of the MAV, IMLEO and cost.

Mars Semi-Direct became the basis for the DRA, which reviewed four options with regard to ascent propellant:

1. No ISPP at all, i.e. all the CH_4 and O_2 required for ascent (about 30 tonnes) is brought from Earth. This option is the least risky, but the most expensive and difficult.

2. About 6.5 tonnes of CH_4 are brought from Earth and 23 tonnes of O_2 are

manufactured locally by reducing atmospheric CO_2. This is a reasonable compromise, as the ISPP system would be very simple, and yet the descent mass would still be greatly reduced.

3. About 2 tonnes of H_2 are brought from Earth, and the methalox is manufactured via the Sabatier and RWGS reactions, as in Mars Direct. This option is technically more complex, but not overly so, and has already been demonstrated by Dr Zubrin and colleagues at Martin Marietta 21 years ago.

4. None of the ascent propellant is brought from Earth; all is made on Mars.

This fourth option is the most challenging, but is favoured in the IMRS plan.

10.3.2. Benefits of 100% ISPP

Producing methalox entirely from indigenous Martian resources is challenging mainly due to the low concentration of atmospheric hydrogen and the difficulty of obtaining hydrogen from the ground. However, although this approach necessitates a more complex ISPP system, there are several important benefits:

1. If methalox is made from H_2 brought from Earth as in Mars Direct, insufficient O_2 is produced for optimal combustion. That's why, in Mars Direct, additional O_2 is produced from CO_2 using the RWGS reaction. By using local H_2O instead of brought H_2, electrolysis of the water produces more than enough O_2, obviating the need for a RWGS subsystem.

2. It develops an essential technological capability that will benefit all future human missions and settlements. Explorers and settlers need water for life support, cooking, cleaning, food production and other things, and hydrogen for manufacturing propellant and other valuable substances. Mars has an abundance of water, and learning how to obtain it is fundamental to settlement.

3. It therefore saves time and money. Rather than investing in the knowledge and technology needed for transportation of hydrogen or methane to Mars, which would only be useful for a few missions, it will be more efficient to proceed directly to the development of knowledge and technology known to be required for long-term settlement, i.e. extraction of water and hydrogen, and propellant production, from local resources.

4. It makes the descent mass of the MAV as light as possible, since the mass and volume of the ISPP system — even a more complex one that includes a mobile robot — is much less than the mass and volume of the propellant. This greatly reduces the difficulty involved in landing the MAV on Mars.

5. As discussed, it may become possible for the MAV to have a common

ascent/descent stage, further reducing its mass and greatly simplifying the design of the vehicle.

Previous architectures have opted to transport none or some of the propellant to Mars as a compromise because of the difficulty of obtaining hydrogen. However, this compromise creates less advantage than it would initially seem, since the reduced complexity of the ISPP system is largely offset by the increased complexity of the EDL and propulsion systems.

In addition, it does not greatly reduce risk. Even though producing propellant for the MAV is in the critical path for the mission, there is never any risk of LOC or LOM (Loss Of Mission) because the crew will only depart Earth if ISPP has already worked as planned, and the MAV is filled with propellant and ready to go. If the ISPP technology does fail, the crew simply waits for a new MAV with a better ISPP subsystem to be built and sent to Mars. The only loss is a delay of the mission until the next launch opportunity, and the cost of ongoing development. Although this would be inconvenient, it is hardly a show-stopper, and much would be learned from the initial MAV/ISPP mission.

With these things in mind, a superior strategy is clearly to embrace the more ambitious but significantly more rewarding goal of manufacturing 100% of the ascent propellant on Mars. Development of this capability is essential and inevitable, and there's no benefit to postponement. A much greater ROI can be achieved by including this requirement from the outset. There's no question that a collaboration of the world's top space engineers, with access to 21st-century tools, can achieve this technological goal.

10.3.3. ISPP Process

The ISPP process for the IMRS MAV is as follows:

Step 1: On arrival at Mars, a door opens in the base of the MAV allowing the AWESOM robot to deploy. It is initially attached to the end of a large roll of PV material, which it unrolls by driving directly away from the MAV, before disconnecting from the end of the roll. The PV blanket provides power to the ISPP system. Additional power may be provided by a PV coating on the MAV's exterior surface, and/or an ASRG, which can additionally provide heat for the electronics and water tank.

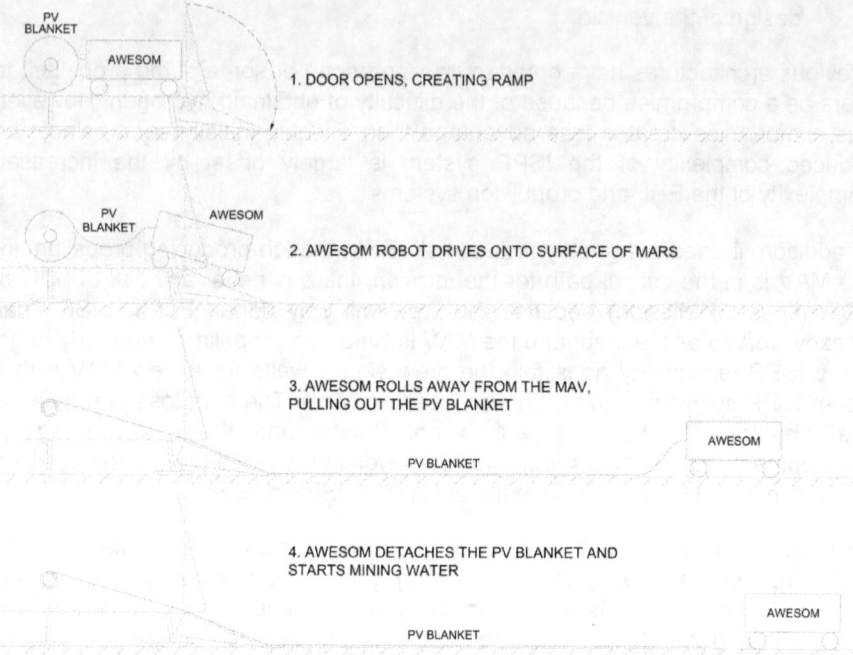

Detail showing rover door operation and PV blanket deployment

Step 2: CO_2 is produced by pumping in Martian atmosphere through a dust filter and drying it via zeolite adsorption. The small amount of water captured is stored in the water tank. The gas is pressurised and allowed to cool to ambient Martian temperatures, causing the CO_2 to condense, and enabling it to be separated from the remaining gases.

Step 3: The AWESOM robot traverses the nearby ground, heating and microwaving the regolith beneath it and collecting the released water. At the end of each sol the robot returns to the MAV to deliver its payload of water by means of a heated hose attached to a robotic arm. The hose nozzle is plugged into a valve in the base of the MAV, and the water is pumped into the MAV's water tank.

Possible track for the AWESOM around the MAV

Step 4: Water is separated via electrolysis into H_2 and O_2:

$$2 H_2O(l) \rightarrow 2 H_2(g) + O_2(g)$$

The O_2 is stored cryogenically as LOX.

Step 5: CH_4 is produced by reacting CO_2 (produced in step 2) with H_2 (produced in step 4), via the Sabatier reaction:

$$CO_2(g) + 4 H_2(g) \rightarrow CH_4(g) + 2 H_2O(v)$$

The Sabatier reaction proceeds at around 600 K in the presence of a catalyst of ruthenium on alumina. This heat could be provided by an RTG, ASRG or an electric heater. H_2O produced by the reaction is stored in the water tank, and is used by the electrolysis unit to produce more H_2 and O_2.

MAV ISPP Schematic

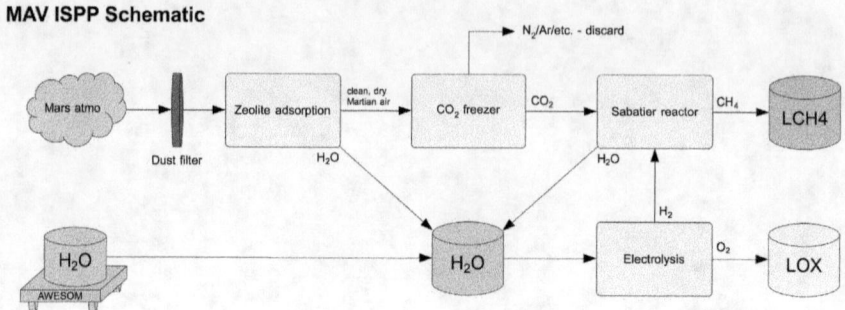

MAV ISPP Schematic

10.3.4. Ascent Mass Estimate

The amount of propellant to make depends on the dry mass of the vehicle. Therefore, it's necessary to make an estimate of this mass.

Kepler

The dry mass of a Dragon V2 capsule is normally 4200 kg. Its heat shield is comprised of 8 cm thick tiles made of a material known as PICA-X (PICA means "Phenolic Impregnated Carbon Ablator". PICA-X is PICA as made by SpaceX). PICA-X is very light, and the total heat shield mass is only about 230 kg. The landing legs, and SuperDraco engines and associated tankage, are estimated to mass approximately 1 tonne.

A modified Dragon without these elements is therefore estimated to mass approximately 3 tonnes. In reality, due to expected improvements in manufacturing and materials, the finished mass will probably be lower.

The Dragon capsule can carry up to 1290 kg of propellant for the Draco engines. This is N_2O_4/MMH (nitrogen tetroxide oxidiser and monomethyl hydrazine propellant), which is storable and non-cryogenic.

Estimated mass: 4290 kg

Superstructure

The mass of the superstructure is crudely estimated from 5 mm thick titanium-carbon composite with a density of about 2800 kg/m^2, giving it an approximate mass of 2200 kg. However, this mass could perhaps be reduced using nanostructured materials.

The superstructure is also made lighter by the fact that it only needs to stand up in Mars gravity, not Earth's. Nonetheless, it must still be capable of withstanding the stresses of descent and ascent.

Estimated mass: 2200 kg

Propellant tanks

The exact mass of tankage will be dependent on the quantity of propellant. Without going into the somewhat convoluted calculations here, the mass of the cryotanks should be about 945 kg. The mass of MLI surrounding the tanks, assumed to be 50 mm thick and having 50 kg/m^3 density, is estimated to be 176 kg.

Estimated mass: 1121 kg

Engines

The estimated TWR for the new engine is 75. This is higher than TWR of the XR-5M15, which is 71.3, but the proposed engines are larger, and, generally speaking, the larger the engine, the better the TWR. This gives an engine mass of about 120 kg each.

Estimated mass: 480 kg

Avionics

The mass of the avionics can be estimated using conventional mass estimating ratios.

Estimated mass: 270 kg

Payload

The payload mass is comprised of six humans (62 kg each) with marssuits (20 kg each) and kit bags (20kg each), plus approximately 300 kg of areological (and hopefully also biological) samples to bring back to Earth.

Estimated mass: 912 kg

Component	Mass (kg)
Kepler + propellant	4290
Superstructure	2200
Propellant tanks + MLI	1121
Engines	480
Avionics	270
Crew + samples	912
Total	**9273**

Because this is all very approximate (and possibly optimistic), and to allow for additional items such as fuel lines and coatings, the total is rounded up to 9.5 tonnes for good measure.

10.3.5. Ascent Propellant Estimate

The propellant mass can be estimated from the ideal rocket equation:

$$\Delta v = v_e \ln(m_i / m_f)$$

Where:

Δv is the change in velocity

v_e is the effective exhaust velocity

m_i is the initial mass

m_f is the final mass

The Δv from Mars surface to the 250 km x 1 sol HEMO is 5625 m/s.

The exhaust velocity (v_e) can be calculated from the I_{sp}:

$$v_e = I_{sp} * g_0$$

Where g_0 is "standard gravity", i.e. gravity at sea level on Earth: 9.80665 m/s^2. This gives:

$$v_e = 380 \text{ s} * 9.80665 \text{ m/s}^2$$

$$= 3727 \text{ m/s}$$

The initial mass can be calculated as follows:

$$m_i \quad = m_f * e^{\Delta v/ve}$$

$$= 9500 \text{ kg} * e^{(5625 \text{ m/s} / 3727 \text{ m/s})}$$

$$= 42{,}972 \text{ kg}$$

The propellant required is therefore:

$$m_p \quad = 42{,}972 \text{ kg} - 9500 \text{ kg}$$

$$= 33{,}472 \text{ kg}$$

The stoichiometric ratio for methalox is 1:3.5, therefore, about 7438 kg of CH_4 and about 26,034 kg of O_2 are required.

This quantity of CH_4 is about 30% of the quantity in Mars Direct (24 tonnes), and about 110% of the quantity in the DRA (6567 kg). The reason for the higher propellant mass compared with the DRA is the higher wet mass (nearly 43 tonnes compared with 40 tonnes). This is because the MAV in the DRA uses a separate descent and ascent stage, which optimises the ascent mass at the cost of a much higher descent mass. This is a trade off. In Blue Dragon, the price of a common descent/ascent stage (which results in a significantly lower landed mass of the MAV and therefore much simpler and easier EDL) is a slightly heavier ascent vehicle. Even so, with new materials and some clever engineering, it may yet be possible to reduce the wet mass of the MAV to about the same value of 40 tonnes.

Estimating the quantity of source reactants

The methalox propellant is manufactured from locally-sourced CO_2 and H_2O. The overall equation is:

$$CO_2 + 2 H_2O \rightarrow CH_4 + 2 O_2$$

Mass ratios can be determined from the atomic mass units of the elements. The atomic mass (in unified atomic mass units, or 'u') of carbon is approximately 12, hydrogen approximately 1, and oxygen approximately 16. The required quantities of each reactant and product are determined by multiplying each mass in u by a factor of 7438 kg/16 u = 465 kg/u.

Molecule	Molecular mass (u)	Quantity	Mass (u)	Mass (t)
Reactants				
CO_2	44	1	44	20,460
H_2O	18	2	36	16,740
Total			**80**	**37,200**
Products				
CH_4	16	1	16	7440
O_2	32	2	64	29,760
Total			**80**	**37,200**

As only about 26 tonnes of O_2 are needed for propellant, the ISPP process will produce a surplus of up to about 3.7 tonnes. This could be used to supplement air supplies for the SHAB or the CAMPER, or for marssuits; however, the ISAP will also produce a surplus of O_2 and will be a more convenient place for replenishing marssuit tanks. In order to minimise tank mass, surplus O_2 can safely be released to the atmosphere (thus contributing, in a small way, to terraforming).

Obtaining CO_2

The Martian atmosphere is approximately 96% CO_2, therefore obtaining 20.5 tonnes of CO_2 requires processing about 21.4 tonnes of atmosphere. The density of the atmosphere at the surface of Mars is approximately 0.015 kg/m^3, so this is about 1.4 cubic kilometres of atmosphere.

If the goal is for the MAV to be filled with propellant within 584 sols, which is roughly the time difference between the arrival of the MAV at Mars and the departure of the crew from Earth, this will require processing about 37 kg (2400 m^3) of atmosphere per sol. Further analysis is necessary to determine how achievable this is using only solar power. If the pump only has power for half of each sol, on average, it must pump about 200 m^3 of atmosphere per hour, which seems pretty ambitious.

If the period is extended to 760 sols, which is the approximate time difference between arrival of the MAV and arrival of the crew (depending on the relative trip times), then the processing rate reduces to about 28 kg/sol.

In actuality, there are 44 months between the arrival and departure of the MAV. An even slower processing rate is therefore possible, and may be acceptable if necessary. However, sending the crew to Mars without the propellant production completed increases risk. If it isn't complete when the crew arrive, they won't be able to move the AWESOM robot, PV material or any ISPP equipment to the SHAB.

Obtaining H_2O

As described in the section about <u>In Situ Water Production</u>, the AWESOM robot can extract an estimated 7.84 kg of water per square metre of ground. The area the AWESOM needs to cover is therefore:

16,740 kg / 7.84 kg/m^2 = 2135 m^2

This is a circle of approximately 26 metres radius.

For the MAV to make all its propellant in 584 sols, the AWESOM must collect about 29 kg of water per sol on average, which requires covering about 3.7 m^2 of terrain per sol. This should be easily achievable.

Because this is a small area, the AWESOM system could be designed to collect somewhat more than this, to allow for variations in water concentration and availability of solar energy, and difficult terrain. A reasonable design goal could be to build the robot to be capable of collecting and carrying up to at least 40-50 kg of water per sol. This will enable the water for ISPP to be collected as early as possible, which is important partly for the peace of mind that will come from knowing the MAV has propellant, but also because H_2 is needed before CH_4 can be produced.

10.3.6. MARCO POLO

The integrated ISRU system within the MAV, including the ISPP and AWESOM subsystems, is similar to the MARCO POLO mission recently proposed within NASA (Interbartolo et al. 2013). Ongoing design and development of the MAV's ISPP features would benefit greatly from the research already conducted in the context of MARCO POLO.

MARCO POLO is a robotic technology demonstration mission comprised of an octagonal lander and mobile robot, which process Martian soil and atmosphere to produce O_2, H_2O and CH_4.

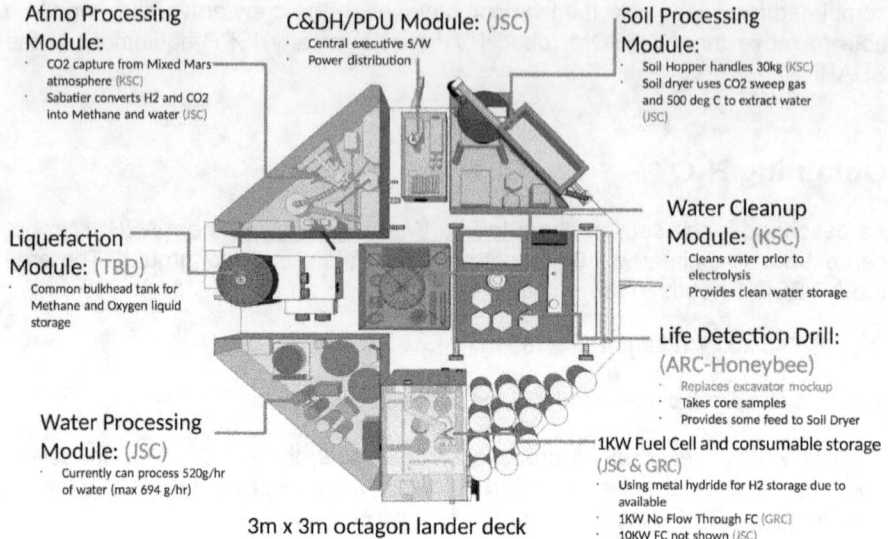

Atmo Processing Module:
- CO2 capture from Mixed Mars atmosphere (KSC)
- Sabatier converts H2 and CO2 into Methane and water (JSC)

C&DH/PDU Module: (JSC)
- Central executive S/W
- Power distribution

Soil Processing Module:
- Soil Hopper handles 30kg (KSC)
- Soil dryer uses CO2 sweep gas and 500 deg C to extract water (JSC)

Liquefaction Module: (TBD)
- Common bulkhead tank for Methane and Oxygen liquid storage

Water Cleanup Module: (KSC)
- Cleans water prior to electrolysis
- Provides clean water storage

Life Detection Drill: (ARC-Honeybee)
- Replaces excavator mockup
- Takes core samples
- Provides some feed to Soil Dryer

Water Processing Module: (JSC)
- Currently can process 520g/hr of water (max 694 g/hr)

3m x 3m octagon lander deck

1KW Fuel Cell and consumable storage (JSC & GRC)
- Using metal hydride for H2 storage due to available
- 1KW No Flow Through FC (GRC)
- 10KW FC not shown (JSC)

Layout of MARCO POLO submodules (Credit: NASA; Interbartolo et al. 2013)

The main differences are as follows:

1. In MARCO POLO, a robot collects dirt and transports it to the lander for water extraction. With AWESOM, the dirt remains in place and water extraction is performed by the robot itself. AWESOM brings the water back to the lander for purification and electrolysis. This is more efficient, and will enable a larger quantity of water to be collected. Most importantly, it enables water to be obtained from permafrost, which would be too difficult to dig up and transport to a lander.

2. MARCO POLO tests closed-loop power production by burning methalox in a fuel cell. In the MAV, the methalox is kept for use as ascent propellant. Power is produced from solar energy and/or an ASRG.

3. The MAV does not include any life detection experiments.

10.4. Returning to Space

10.4.1. Ascent

At some point after the completion of propellant production and prior to the launch of the MAV, the crew remove the external panels around the base of the MAV, unbolt and extract the ISPP modules, and relocate these, along with the AWESOM rover and PV blanket, to the SHAB.

Prior to launch, the areological samples are transferred to *Kepler*.

At departure time, the crew pile into the CAMPER, wearing their marssuits and carrying their kit bags. They drive to the MAV and transfer into *Kepler*. The CAMPER is then driven back to the SHAB via remote control.

The MAV is launched when *Adeona* is close to periareon. Its propellant tanks are initially filled with about 33.5 tonnes of methalox. The engines propel it upwards from the surface of Mars, with a trajectory curving over until the MAV is flying parallel to the surface of Mars in the same 250 km x 1 sol orbit as *Adeona*.

Acceleration estimate

There are two main factors to consider when designing for ascent engine power:

- The local gravity, which the engines must overcome in order to reach orbit.

- Limits on g-forces that can be withstood by humans. This second factor is checked below.

The local gravity on Mars is:

g_M = 3.71 m/s^2

The initial, wet mass of the MAV is:

m_i = 43 t

Thus the force of its weight on Mars is:

F_W = $m_i \times g_M$

= 160 kN

The engines must produce thrust greater than this weight in order to generate upward movement. The descent/ascent engines provide 90 kN each; thus, the total force produced by four engines is 360 kN. The net initial upward force on the MAV is therefore:

F = 360 kN - 160 kN

= 200 kN

The net upward acceleration is:

a = F / m

= 200 kN / 43 t

$$= 4.65 \text{ m/s}^2$$

The total initial g-force experienced by the crew will be:

$$g_i \quad = 3.71 \text{ m/s}^2 + 4.65 \text{ m/s}^2$$

$$= 8.36 \text{ m/s}^2$$

$$= 0.85 \text{ g}$$

This is slightly less than Earth gravity, although it will be the highest gravity level the crew has experienced for a while. The g-forces will steadily increase as the MAV climbs to orbit and propellant is consumed.

Once on orbit, the downward force on the MAV is balanced by the centripetal force produced by its circular path, thus producing zero g-forces in the vertical direction. The axial force, however, will be the full 360 kN produced by the engines.

The maximum acceleration experienced by the crew will be right at the end of the burn, when the vehicle is at its lightest. When the MAV reaches *Adeona*'s orbit its tanks will be empty, thus its final mass will be its dry mass of 9.5 tonnes. The maximum acceleration is therefore:

$$a \quad = F / m$$

$$= 360 \text{ kN} / 9.5 \text{ t}$$

$$= 37.9 \text{ m/s}^2$$

$$= 3.86 \text{ g}$$

This is an acceptable g-force for trained astronauts. As mentioned earlier, acceleration forces are capped at 4 g's for crew health and comfort, and so they can reach the controls. Space Shuttle crews, by comparison, experienced a maximum of 3 g; but let's assume Mars crews are made of tougher stuff. Humans can tolerate 5-6 g for about 10 minutes.

10.4.2. Mars Orbit Rendezvous

Adeona will be in HEMO with periareon at 250 km and a period of 1 sol, inclined at an angle matching the latitude of the IMRS. It will therefore pass directly over the IMRS once per day. The timing and trajectory of the MAV's launch must be as accurate as possible such that it coincides with *Adeona* passing overhead, and so the MAV ends up reasonably close to *Adeona* and on a matching trajectory.

Kepler will then separate from the main body of the MAV.

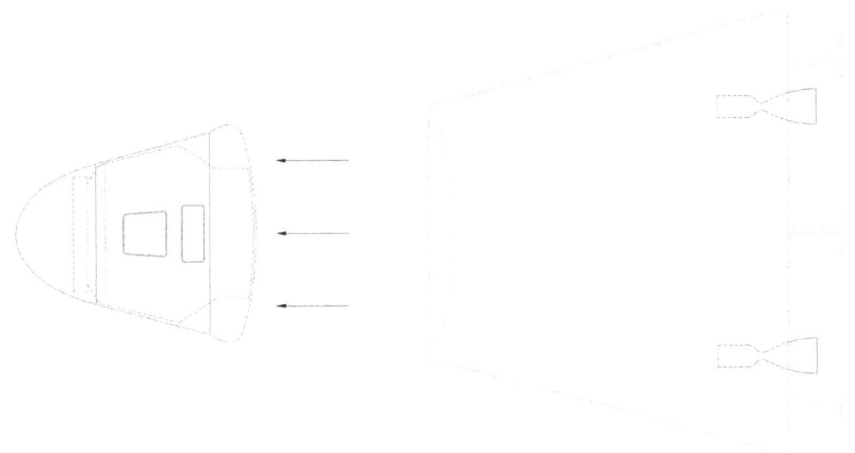

The main body of the MAV will orbit Mars for a while before falling through Mars's atmosphere back to the surface.

Kepler's nose cap will open, revealing the docking port.

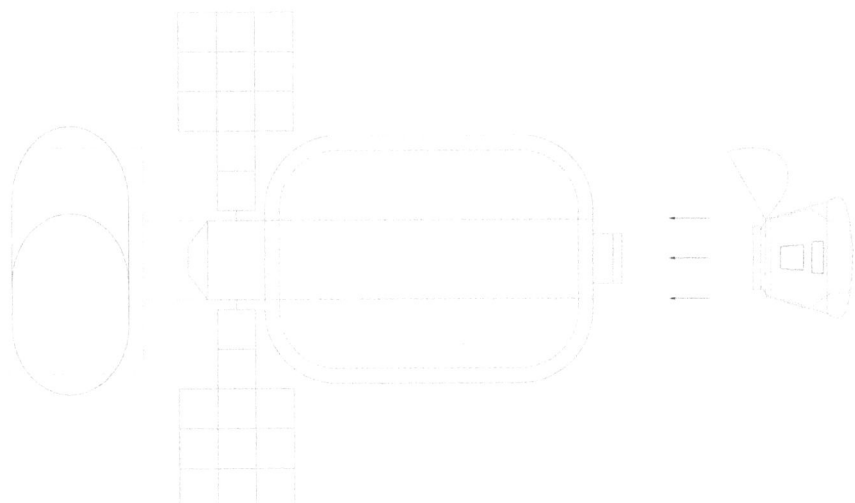

Kepler will use its Draco engines to line itself up with the docking port on the end of the THAB, and dock.

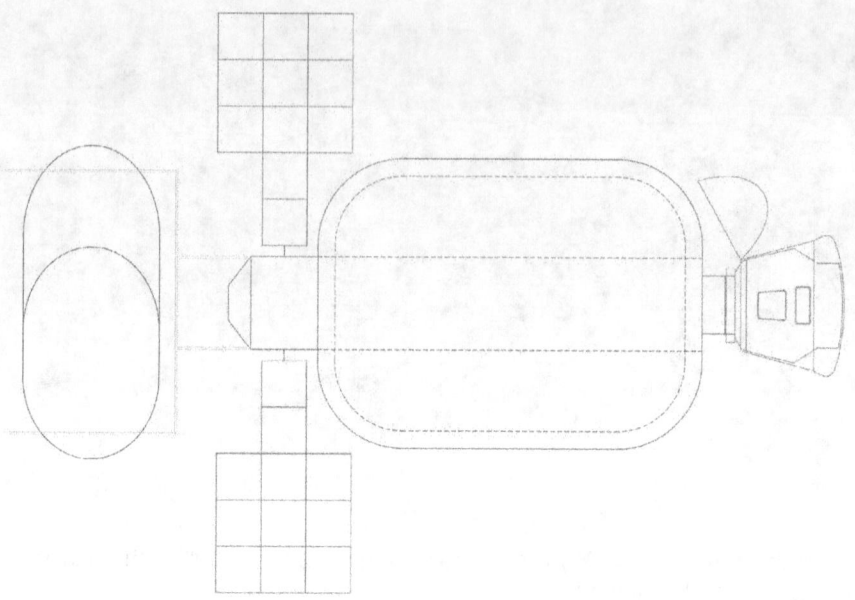

Once all samples have been transferred from the capsule into the THAB and safely stored, the THAB's hatch will be closed, and *Kepler* jettisoned. It will orbit Mars for a while before burning up in its atmosphere.

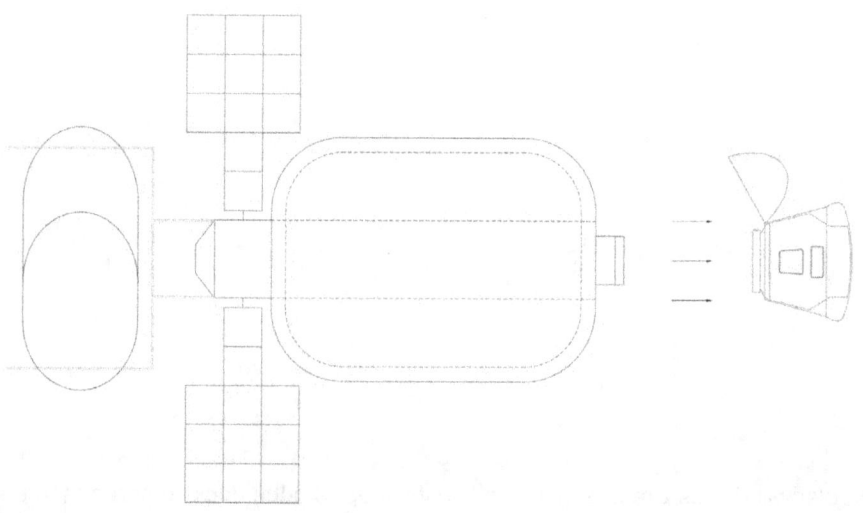

Breathing Martian air

After docking, and before the forward hatch opens, the atmospheres in the THAB and *Kepler* will be very different. Inside the THAB will be a basically Earthian atmosphere, about 79% N_2 and 21% O_2 at 101.3 kilopascals pressure and 298 K, as the ECLSS in the THAB will have been remotely activated in advance to warm and freshen the environment. The capsule will still contain Martian air, however: 96% CO_2, 2% N_2 and 2% Ar at only 600 Pascals. Although this air would have been a chilly 210 K (or thereabouts) at launch, it will have warmed somewhat during the short trip to orbit due to heat from the marssuits and the spacecraft's systems.

If the astronauts open the forward hatch directly, the rapid expansion of air into the capsule could cause injury. Therefore, the forward hatch will be fitted with a release valve that will allow the pressure to equalise gradually. Once the pressure in the capsule has equalised with the pressure in the THAB, it will then be safe for the crew to open the forward hatch and transfer into the THAB, where they can clean any dust off their marssuits and remove and store them.

The Martian air from *Kepler* that will mix with the Earthian atmosphere of the THAB will not pose a toxic hazard. The capsule's pressurised volume is 10 m^3 and the THAB's is 330 m^3; however, the atmospheric pressure in the THAB will be approximately 170 times greater than that in the capsule. The mass of the atmospheric gases in the THAB is approximately 390 kg, but the mass of gases in the capsule is a mere 150 grams. The concentrations of problematic gases such as CO, NO and O_3 that will be introduced into the THAB will therefore be very small once distributed throughout the THAB, and the ECLSS will mop them up pretty quickly anyway. The 144 grams of Martian CO_2 added to THAB, which will be virtually devoid of CO_2 due to having been uninhabited for the past 1.5 years, is less than the usual operating amount of 200 grams.

The loss of *Kepler*

Jettisoning *Kepler* may seem like a waste of a capsule. Another approach would have been for *Kepler* to remain docked to *Adeona* for the return trip and used for landing on Earth, much like *Einstein* on the outbound trip. However, it's preferable to discard the capsule for three important reasons:

1. Planetary protection. *Kepler* has been sitting on Mars for 44 months and could be contaminated with Martian organisms. Although it's considered unlikely that biological life could survive on the surface of Mars due to the high UV radiation and extreme dryness, there is still a risk that unknown organisms could attach themselves to the capsule, survive the journey to Earth, and enter Earth's environment. COSPAR (Committee on Space Research) recommendations with regard to planetary protection would therefore be against landing this capsule on Earth.

2. If *Kepler* is not needed for an Earth landing its mass can be substantially reduced, as no heat shield, thrusters or propellant are required for EDL. This reduces the mass to be landed on, and launched from, Mars, saving

considerable expense and difficulty.

3. Without *Kepler* docked to *Adeona*, much less propellant is required for TEI.

Discarding *Kepler* incurs an additional expense of launching a Falcon 9 with *Newton*. However, as these are reusable items, the only cost is for the launch and propellant.

A future MAV could perhaps be made fully reusable, with both the lower section and the capsule landing back at the IMRS, where they could be reused, repurposed for storage or shelter, or salvaged for parts. Another approach could be to keep *Kepler* on orbit for later recovery and reuse as an orbital transfer vehicle. However, this would require a supply of propellant for station-keeping.

11. Operational Elements

This section describes additional hardware elements involved in the mission architecture and development of the IMRS.

11.1. Mars Surface Habitat

The purpose of the Mars Surface Habitat (SHAB) is to healthfully accommodate crews of up to six astronauts at a time on the surface of Mars for about 1.5 years. It must incorporate sleeping quarters, food preparation facilities, workstations, laboratories and other work areas, bathroom, medical facilities, subsystems for ECLSS, power, communications and computing, an airlock, and storage for food, water, air, marssuits, personal effects and other items.

As the THAB is based on a Bigelow Aerospace B330 optimised for interplanetary space, so the SHAB is based on a B330 optimised for use on the surface of Mars. The B330 modules provide radiation protection comparable to the ISS, which will protect the crew during their stay and obviate the need to cover the habitat in dirt to provide additional protection (a process often recommended for Mars habitats, but one that would be challenging). A B330 module is well-suited for this purpose, yet should be much cheaper than a custom-built unit, being a COTS item, although some customisation is inevitable.

An important advantage is gained by using the same habitat technology — the B330 — for both the THAB and the SHAB; namely, commonality of hardware. It means there will be fewer systems for the crew to learn and understand. If they can repair the THAB they'll also be able to repair the SHAB, should the need arise.

11.1.1. Entry, Descent and Landing

A B330 module weighs approximately 20-23 tonnes, which necessitates a SHLLV in order to deliver one to the surface of Mars. Reducing this mass would almost certainly require reducing the crew size, simply because of the amount of equipment and supplies necessary to keep people in good health for this length of time. Most estimates for Mars habitats exceed 20 tonnes. The B330 is a comparatively light option yet providing ample living volume.

Blue Dragon, like most Mars mission architectures, is reliant on the availability of a suitable SHLLV. It's possible there will be a choice of several such vehicles available in the desired time frame:

1. Space Launch System, currently being developed by NASA.

2. Long March 9, being studied by CNSA.

3. The new megalauncher planned by Roscosmos.

4. Mars Colonial Transport in development at SpaceX.

For now, the SLS 130t Cargo is considered the most suitable near-term vehicle for delivering payloads of 20-30 tonnes to the surface of Mars, being the furthest along in its development. If an SLS can deliver 30 tonnes to Mars, a mass budget of about 7-10 tonnes of supplies and equipment may be sent along with the habitat. Because the total quantity of supplies and equipment required by a crew will likely exceed this amount, additional tonnage would be delivered with the Red Dragon "Mars Supply Capsules".

After the first couple of missions, when SHABs are being reused by subsequent crews, ongoing resupply may also be achieved using capsules, or newer vehicles such as the MCT.

11.1.2. Orientation

For the basic form of the SHAB, previous architectures such as Mars Direct have primarily focused on a squat, two-or three-floor vertical cylinder, or the so-called "tuna can" model. This has become the traditional de-facto design:

"Tuna can" Mars habitats (Credit: The Mars Society)

These habitats nominally have a diameter of around 8 metres, which fits neatly inside the 8.4 metre diameter payload fairing of an SLS rocket.

One notable exception is the MARS-Oz habitat design:

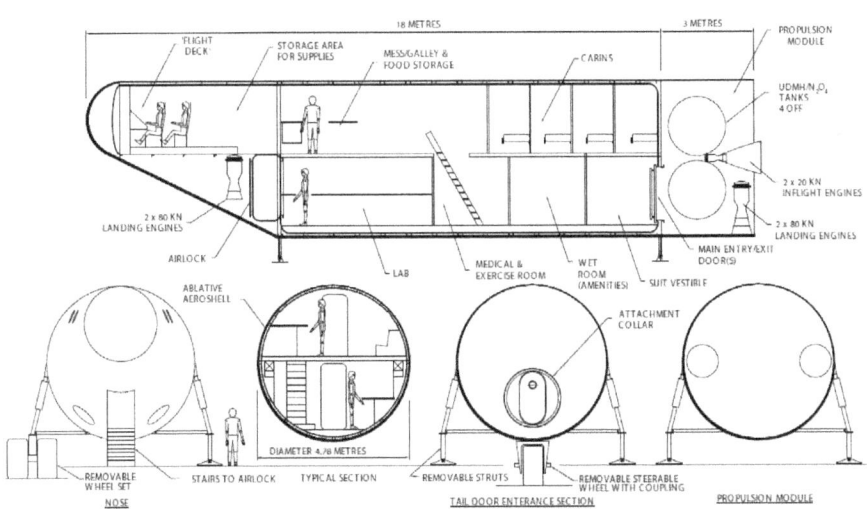

MARS-Oz habitat (Credit: Mars Society Australia)

Although a B330 may seem ideal for vertical orientation like the tuna-can design, because the airlock is at the end, it is more useful if oriented horizontally. This model of a Mars base constructed from B330 modules illustrates a horizontal configuration for the modules:

Mars base comprised of B330 modules (Credit: Bigelow Aerospace)

11.1.3. Initialisation

The SHAB will be sent to Mars at the same time as the MAV, and possibly travel on a slow 9-month trajectory like the supply capsules. The SHAB will not land too close to the MAV, but approximately several hundred metres distant so that the launch of the MAV does not adversely affect any of the SHAB's systems; for example, by overheating anything with rocket exhaust, or covering the SHAB's solar panels in dust.

Once successfully landed on Mars, the SHAB will be remotely activated and several processes initiated:

- Rolls of PV material will be extracted and unrolled with the help of the CAMPER or other mobile robot, and the power system activated.

- The ISAP unit will be activated, and production of O_2 and buffer gas will commence. If the SHAB is not delivered in a fully inflated state, then it will be inflated using this air.

- Water tanks will begin to be filled by the ISWP system.

11.1.4. Redundancy and Safety

One of the advantages of Blue Dragon is that, theoretically, only one habitat needs to be delivered to Mars, which can then be used for multiple missions.

In practice, however, a minimum of two habitats will be preferable in order to provide one or more backups. Blue Dragon provides every crew with a backup SHAB from the very first mission. The first SHAB (SHAB-1) is sent in the predeployment phase for the Alfa Mission, scheduled for 2031, and will therefore have been activated, tested and filled with air and water in preparation for the arrival of Alfa Crew approximately 2 years after. Note, however, that the launch opportunity for the Alfa Crew outbound trip in 2033 is the same as for the predeployment phase of Bravo Mission. Hence, the second SHAB (SHAB-2) will be en route to Mars at the same time as Alfa Crew, along with the second MAV and other equipment required for Bravo Mission.

Most of the equipment being predeployed for Bravo Mission will be travelling on a slower, 9-month trajectory, to conserve propellant, while Alfa Crew will be travelling on a quicker 6-month trajectory, to protect their health. However, SHAB-2 could also be sent on a fast trajectory in order to ensure that Alfa Crew has a backup SHAB. When SHAB-2 arrives at the IMRS, they can monitor its arrival and setup. Once its systems are fully operational, Alfa Crew will have an alternate home in SHAB-2 should there be any malfunction with SHAB-1 during their surface mission. Assuming that the two SHABs are well-maintained during their lifetime, subsequent missions will have a spare SHAB available if necessary.

Another approach would be to send SHAB-1 and SHAB-2 to Mars at the same time, and Alfa Crew can ensure that both are fully operational in preparation for future crews.

Assuming that the crew size for IMRS missions remains constant at six, only two SHABs are required for the IMRS: one for the crew, and one spare. The cash saved by not having to predeploy an SHAB from the third mission onwards can be funnelled into other base hardware such as greenhouses, power systems, heavy machinery and other components.

Having said that, the intention is that the base will expand as space technology improves and its cost decreases, participating space agencies seek to expand their activities on Mars, and some astronauts opt to remain for multiple seasons at a stretch or even permanently on Mars. It may be desirable to construct MTVs like Athena with two B330 modules (or perhaps one B2100 module) in order to support larger crews. This will require a matching habitat capacity on the surface.

In lieu of a backup SHAB, the CAMPER, which has capacity for six, may also be used as an emergency lifeboat if necessary. If both SHABs become unusable, the crew can use the CAMPER to reach the MAV and launch and transfer to Adeona, which will have contingency supplies of food and water. There they can wait until the next opportunity to head home.

11.2. Surface Vehicles

The intention is that a single payload of surface vehicles be delivered to Mars using a Falcon Heavy, which is capable of delivering an estimated 13 tonnes to the surface of Mars. This payload would include:

- The CAMPER: a large pressurised robotic vehicle with a 2-person airlock and capacity for six astronauts.

- Attachments for the CAMPER such as a drill, excavator bucket, and trailer.

- Three unpressurised ATVs (All-Terrain Vehicle).

11.2.1. Crewed Adaptable Multipurpose Pressurised Exploration Rover

The architecture calls for a highly capable, multipurpose pressurised surface vehicle, capable of operation in both crewed and autonomous modes. This versatile vehicle will be central to the surface mission from beginning to end.

The CAMPER will fulfil a variety of functions:

1. On arrival, the CAMPER will explore the surrounding environment finding safe pathways between base components, namely the SHAB, *Einstein*'s LZ, and the MAV.

2. The CAMPER may assist the MAV and the SHAB in deployment of their photovoltaic blankets (unless this can be more effectively achieved using the AWESOM robot, a crew member, or other method).

3. The CAMPER will pick up the crew from *Einstein*'s LZ when they first arrive, then transport them to the SHAB.

4. If an excavator attachment is included, the CAMPER can clear makeshift roads between different areas of the base.

5. The CAMPER will be used on a regular basis for exploring the surrounding territory, both for short, local sojourns, and longer, overnight or multisol excursions.

6. At the end of the surface mission, the CAMPER will transport the crew and their samples from the SHAB to the MAV.

7. It can be used as an emergency "lifeboat" in the event of an ECLSS failure in the SHAB (if a backup SHAB is unavailable).

Design

Rather than presenting a completed design, this section simply offers a handful of ideas that may influence a future design study of the CAMPER. Many designs have been proposed for Mars pressurised surface vehicles, which give an indication of what the CAMPER may look like.

The Manchu Rover, shown below, is probably somewhat larger than the CAMPER needs to be.

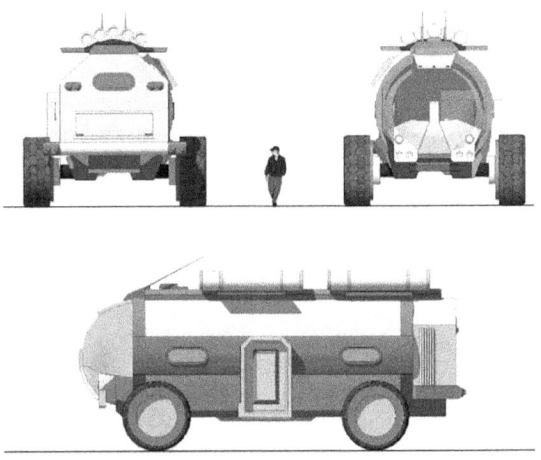

Manchu Rover (Credit: Da Vinci Mars Design)

In contrast, the "Marsupial Rover" designed by Graham Mann and David Willson from MSA (Mars Society Australia), shown below is possibly somewhat too small, as the CAMPER must be able to seat a full crew of six. The Marsupial Rover is part of the MARS-Oz architecture developed by Mann, Willson and Jonathan Clarke, and manufacture of a prototype is almost complete.

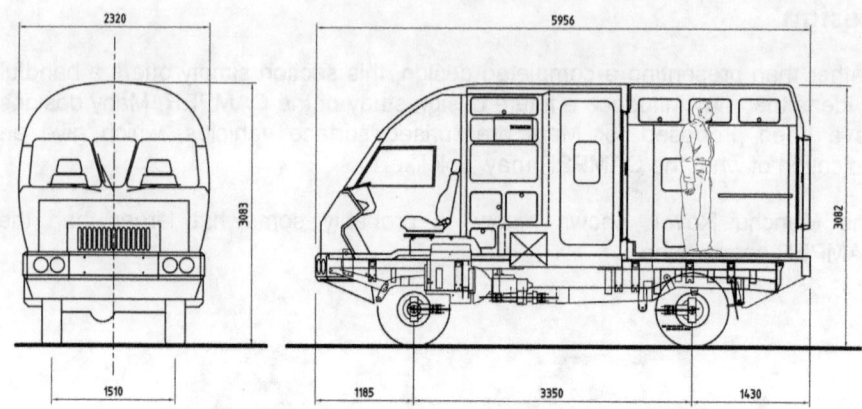

Marsupial Rover (Credit: Mars Society Australia)

Below is shown NASA's current prototype for a lunar vehicle known as the "small pressurized rover", which will no doubt inform future designs for a Mars pressurised vehicle.

Small Pressurized Rover (Credit: NASA)

Capacity

Since the CAMPER picks up the whole crew from *Einstein*'s LZ on their arrival, and again transport the whole crew to the MAV for their departure, it must have capacity for the entire crew of six. Because standing may be difficult after spending months in microgravity, it is necessary to provide some kind of seating for all of them.

However, carrying six will be the exception rather than the norm. A typical excursion will only involve two or three crew members, and the spare seats can be converted into beds for longer excursions.

Communications

The CAMPER will constantly report its current position and condition to the SHAB, where the control room is located. This information will include the internal atmospheric pressure and constituents, internal temperature, and supplies of oxygen, buffer gas, water, food, and propellant and/or electrical energy. In this way the CAMPER can be monitored remotely, whether occupied or not. Astronauts at the SHAB will have real-time awareness of its condition at all times, including while other crew members are out exploring.

Power system

The CAMPER will most likely be solar-electric, with electric motors and high-efficiency PV panels on the roof. The vehicle will be fitted with an electricity regulator that will preferentially use solar energy, with surplus energy being stored in batteries or graphene supercapacitors.

Electric vehicle technology is currently receiving considerable investment by the automobile industry, and advancing rapidly. Battery technology in particular has improved considerably in recent years, especially since the emergence of Tesla Motors.

Batteries or supercapacitors will provide additional power when necessary, for example, at times of low solar energy (i.e. at dusk or at times of high dust), or when the excavator bucket is attached, heavy lifting is required, transporting cargo or navigating difficult terrain. When the CAMPER is not in use, solar energy will charge the batteries or supercapacitors. Although the CAMPER should not be used in low light levels purely for safety reasons, it should nonetheless be able to operate from 100% battery power if necessary.

Alternatively, or additionally, the CAMPER may have a methalox combustion engine. As described in *Case for Mars* (Zubrin & Wagner, 1996), combustion-powered surface vehicles may be the most effective way to cover long distances on Mars and thereby explore a larger area around the base. Methalox is the best propellant for combustion engines on Mars due to its high energy-mass ratio, ease of storage and transportation, and the fact that it can be manufactured from local resources.

If this option is taken, the MAV could be designed to produce a surplus of propellant that could be used by the CAMPER, as in Mars Direct. This would require the MAV to have larger tanks, which would increase its mass, and might compromise the common descent-ascent stage design described earlier. Another approach would be for the CAMPER to have a built-in, smaller-scale ISPP system, but then the obvious question is how it would be powered. Yet another approach would be for the SHAB to include an ISPP system, similar to the MAV's, which would manufacture methalox for the CAMPER.

Airlock and dust mitigation

Unlike some modern pressurised Mars rover designs, the CAMPER will not have suitports, for several reasons:

1. The intention is to use MCP spacesuits, which are not compatible with suitports.

2. Gas pressurised suits with suitports would be too bulky to wear inside the Dragon capsules.

3. The CAMPER must have capacity for the entire crew and it would be impractical to include six suitports.

Instead, the CAMPER will have an airlock. In the interests of efficiently balancing time and volume, this airlock will accommodate up to two people at once, as this will be the most common excursion crew size, and crew members can help clean each other's suits during ingress (see below). For the whole crew to enter or exit, as is necessary at the beginning and end of the surface mission, the airlock will therefore need to be cycled three times to transfer all six crew members.

For more on this topic, refer to the sections below about Marssuits and Airlocks.

Marssuit support

The intention is that the astronauts will be able to wear their marssuits inside the vehicle, since they will need to be wearing them before entering and after exiting the vehicle. However, the CAMPER will also include racks along one wall for hanging up PLSS backpacks and helmets, and sockets for plugging in suit batteries for recharging from the CAMPER's power system.

The CAMPER will be capable of supporting extended, multi-sol excursions to more distant locations. Storage for marssuits as well as casual clothes will be provided.

Robotic arms and manipulators

The CAMPER is actually a robot, and will have two powerful arms, one on each side, which can be controlled from within the vehicle, from the SHAB, or remotely by MCC on Earth.

The arms will have grippers that may be used for several purposes:

- Picking up samples of dust and rock, and placing these in sample boxes along the sides of the CAMPER. This will enable crew members to collect samples without leaving the vehicle.

- Picking up heavy pieces of equipment or cargo for moving. For example, carrying cargo from supply capsules back to the SHAB.

- Gripping and moving an excavator bucket.

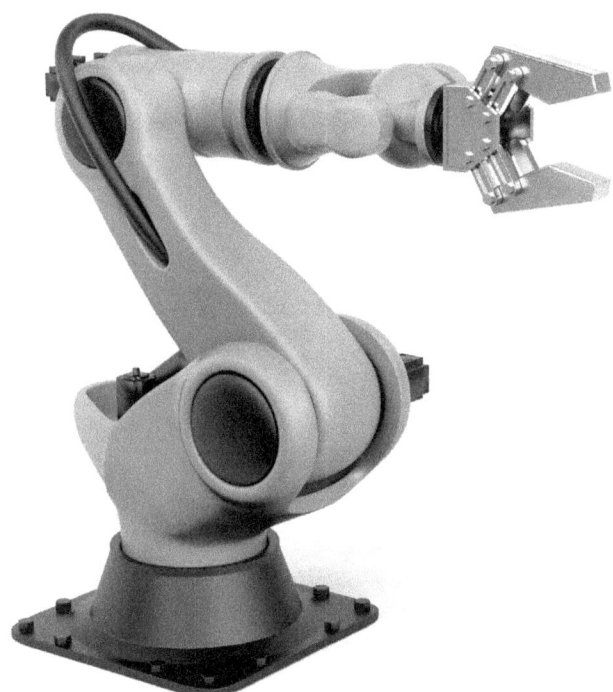

Industrial robot arm (Credit: Kuka Robotics)

Excavator capability

An excavation attachment may be included for clearing makeshift roads, piling loose regolith on a habitat for additional radiation protection and thermal insulation (not that that should be necessary with the B330 modules), or uncovering deeper layers of regolith for closer inspection.

This equipment should be easily attached or detached. A bucket similar to that found on an excavator or front-end loader could be fitted with handles that can be gripped by the grippers on the CAMPER's arms.

One issue with this feature is that the CAMPER needs to remain stable while using it. If it's too light in the back end then carrying the bucket may cause it to tip forwards, particularly when loaded. This problem may be addressed by distributing the bulk of the vehicle's mass (for example, battery packs and ECLSS) towards the rear of the vehicle, as is customary with ordinary bobcats and front-end loaders. Another approach is to use tracks instead of wheels.

Front-end loader (Credit: Sinoway Industrial (Shanghai))

Tracks

In order to provide greater stability for the rover when using the excavator attachment or the drill, tank tracks may be preferable to wheels.

"The Chariot" from Lost in Space (Credit: Fox Television Studios)

Tracks are especially good for traction and will help to prevent the CAMPER from becoming stuck in soft soil, as happened with *Spirit*.

Drill

The CAMPER may also be fitted with a drill suitable for taking core samples. This feature supports the science objectives of the mission, including the search for water, volatiles and biological life, and the investigation of subsurface mineralogy.

Mobile drill rig (Credit: Atlas Copco)

The purpose and design of the drill are based on the core objectives of the Ice Dragon mission (Stoker et al. 2012), which proposes to land a Red Dragon capsule that contains within it a drill for taking core samples (see SpaceX Dragon Capsule). Although the Ice Dragon has not yet been approved, its scientific and engineering objectives remain relevant and practical, and can be incorporated into one or more IMRS missions by the inclusion of a drill capable of penetrating at least two metres into the Martian permafrost.

Aside from investigating subsurface water and organics, a drill will reveal information about the availability of elements in Mars's near-surface crust such as metals and the elements necessary for plant growth. This is important for settlement, as metals are valuable for manufacturing and construction, and plant nutrients are necessary for food production and terraforming. Investigation of the subsurface environment will also provide interesting insights into Mars's areological history.

Available hardware

As one of the cost-reduction strategies employed in the IMRS plan is to make use of COTS hardware where possible, it's worthwhile to investigate if any vehicles already exist that could be a suitable prototype for the CAMPER.

One such vehicle could be the Viking BsV 10 All Terrain Vehicle (Protected) developed for the Royal Marines by Hägglunds Vehicle AB and the UK Ministry of Defence. The Viking is an amphibious tracked vehicle with a road speed of 65 km/h and a water speed of 5 km/h. Each of the four rubber tracks are driven by the powerful diesel engine, enabling it to effectively traverse all types of terrain.

Viking BvS 10 ATV (Protected) (Credit: BAE Systems)

The Viking design includes several features that will be useful on Mars. Although a Mars rover doesn't need to be amphibious, this aspect of the Viking's design implies that the vehicle is at least somewhat airtight, which will be important for the pressurised CAMPER. It's also very tough and mobile, having been designed for use in jungle, desert and polar conditions. With an operating temperature range from -46°C to 49°C the Viking therefore includes heaters, which will also be an essential feature of the CAMPER, although slightly more powerful heaters will be necessary at the IMRS where the temperature will range between about -60°C and 0°C. The Viking naturally also includes integrated communications and electronics.

The vehicle has two separate sections. The front section carries the crew, with the rear section being used differently depending on the vehicle's purpose. For example, the Troop Carrying Variant can carry up to 12 people. However, a better prototype for the CAMPER would be the Repair and Recovery Variant, in which the rear section carries a crane, mobile workshop, air compressor and capstan. In the CAMPER, the front section would be designed to carry up to six astronauts in the front section, plus the airlock and ECLSS. The rear section could be used for the drill rig, rock boxes, mobile laboratory, supplies and other things. It may be a useful feature of the CAMPER for the rear section to be detachable when not in use.

The Viking is powered by a diesel engine, which would be replaced by batteries and/or a methalox engine. The roof could be fitted with solar panels to recharge batteries and provide additional power to electronics, communications, the ECLSS and internal lighting. Although the Viking can travel up to 65 km/h on roads, a maximum speed of 10-20 km/h will be more than enough on Mars. Efficiency and safety will be much more important design goals than speed.

Ultimately something like a Viking may only serve as inspiration for the CAMPER's design rather than a material prototype. Although it may seem cheaper to adapt an existing vehicle, the CAMPER will probably be sufficiently different that constructing an entirely new vehicle will produce a better result, especially since it will need to be built from a much lighter material than the heavy steel used in the Viking, in order to keep the mass low. However, as a large part of development cost is design, starting from an existing proven design will still produce a tangible cost saving. In addition, several subsystems (drive train, communications, controls, display, etc.) may be directly transplantable from a Viking into the CAMPER.

11.2.2. All Terrain Vehicles

Several (2-3) ATVs or UTVs (Utility Task Vehicles) will be included in the equipment predeployed to Mars for Alfa Mission. Future missions may include additional ATV/UTVs or spare parts, depending on requirements.

Each ATV will be capable of carrying 1-2 people in addition to lightweight loads such as tools, samples, supplies and experiments.

Two-person UTV with safety roll bar and small utility tray (Credit: Kawasaki)

The primary purpose of ATVs is to provide convenient and accessible short-range transport for small teams (typically 2-3 people) around the region close to the base. Such uses could include:

- Brief/nearby EVAs.

- Driving between the SHAB and the MAV or Dragon capsules, or other areas of the base, to make inspections or retrieve equipment or supplies.

- Backup surface transport for emergencies.

Their advantages include:

- Low mass.

- Low energy/fuel requirements.

- Faster than walking.

- Less time delay in entering/exiting the vehicle, compared with the CAMPER, due to not having to cycle an airlock.

- Easy mount/dismount for field work, e.g. close inspection of surface features or equipment, or deployment/recovery of surface experiments.

- Capability to haul or carry light/medium-mass payloads.

ATVs may be battery or solar-powered, or they may run off methalox if the MAV's ISPP system is configured to produce a surplus (Zubrin, Baker & Gwynne 1991).

Single-person ATVs have been used as surface vehicles in Mars analogue research at FMARS and MDRS since their construction, and have proven to be exceptionally useful.

Robert Zubrin, Vladimir Plester, and Katy Quinn of FMARS Crew 2 (Credit: The Mars Society)

A convoy of ATVs at the MDRS (Credit: Austrian Space Forum)

The Apollo 15 Lunar Rover bore a closer resemblance to an ATV than the pressurised vehicles now being considered for the Moon and Mars. It may be that ATVs developed for Mars will resemble the Lunar Rover, particularly if designed to accommodate two astronauts in marssuits.

The Apollo 15 Lunar Rover (Credit: NASA)

11.3. Communications

Communications are fundamental to the success of IMRS. The astronauts need to be in contact with Earth, and with each other, 100% of the time. In addition, telemetry data from all spacecraft and base elements must be accessible to MCC on Earth as well as the crew on Mars.

An important consideration when designing communications solutions for HMMs is that the bandwidth requirement will be much higher than has ever been required by a robotic mission, including:

- text, voice and video messaging;

- uploading images and streaming video;

- uploading telemetry data from all capsules, vehicles, habitats and marssuits;

- uploading health data (human telemetry);

- uploading data collected during field work and from experiments;

- downloading websites, music and videos for caching on the SHAB and THAB's proxy servers (although most of this can be done in advance).

To support the high bandwidth requirement, all in-space communications required by the mission will probably utilise lasers. Laser communication in space has been successfully demonstrated by NASA's OPALS (Optical Payload for Lasercomm Science) and LLCD (Lunar Laser Communication Demonstration) missions, and ESA's EDRS (European Data Relay System). OPALS and LLCD set new records for space communications data rates of 50 Mbps and 622 Mbps respectively. The EDRS "SpaceDataHighway" can support 1.8 Gbps over 45,000 km links, potentially increasing to 7.2 Gbps in the near future. Coherent bidirectional space-ground laser communication links of 5.6 Gbps have already been demonstrated.

NASA's LCRD (Laser Communications Relay Demonstration) mission is designed for laser communications between Earth's surface and space, and all around the Solar System. With the first satellite launching in 2017, this mission could be an important precursor to development of an Earth-Mars laser communications network.

11.3.1. Between Earth and IMRS

The IMRS will be able to communicate directly with Earth whenever there's a direct line of sight between them. However, this is not always the case, for two reasons:

Problem 1 - Mars rotates: Because Mars is rotating on its axis, it comes between Earth and IMRS during the Martian night, blocking direct communications.

Problem 2 - solar conjunction: Approximately every 780 days Earth and Mars are on opposite sides of the Sun (Mars solar conjunction). Thermal noise from the Sun prevents direct Earth-Mars communications for about 2-4 weeks. This would occur once during each surface mission.

With no CRSes (Communications Relay Satellite), IMRS would therefore only be able to communicate with Earth about 48% of the time. This may be acceptable for robotic missions, but certainly not for human missions. People on Mars will want and need to be connected with Earth 100% of the time; not just for safety, but also psychological reasons. Back on Earth, friends, family, MCC and the media will all be depending on constant communications with the crew. Any disruption would produce stress on both planets and would greatly amplify the danger of an emergency situation on Mars.

Orbiters

The approach currently taken by robots and spacecraft on the surface of Mars to address the Mars rotation problem is to relay messages through orbiters like the 2001 Mars Odyssey, the Mars Global Surveyor and the Mars Reconnaissance Orbiter. These spacecraft are designed to double as CRSes and are much closer to the surface of Mars than Earth is, which means the signal strength required to communicate with them is significantly less. Being in orbit, they are also in sight of Earth for much longer periods of time.

This approach could be taken by the IMRS; however, there would still be times when no orbiters will be visible from the base, which would interrupt communications with Earth. Besides, the bandwidth provided by existing orbiters is far less than what will be required by the IMRS.

MCOS (Mars Communications and Observation Satellite)

The Mars rotation problem can be almost fully addressed by placing a dedicated satellite in an areostationary orbit above the base. This is a circular orbit at an altitude of approximately 17,031 km above Mars's equator, where it's orbital period matches the period of Mars's rotation, thus causing it to appear stationary with respect to the surface of Mars.

MCOS, which will be a combination of high-bandwidth LCRS (Laser Communications Relay Satellite) and high-resolution Mars observation satellite, would thus be positioned above Mars's equator, at an equal or similar longitude to IMRS.

During the night at IMRS, when the line of sight between IMRS and Earth is blocked by Mars itself, communications will be relayed through MCOS. Because of the high altitude of the satellite (~5 times the radius of Mars), and the inclination of Mars's axis, this capability enables communications to be relayed from IMRS towards Earth about 99% of the time. There will be occasions when MCOS, Mars and Earth will line up and Mars will block the line of sight between MCOS and Earth, but such periods will be infrequent and brief.

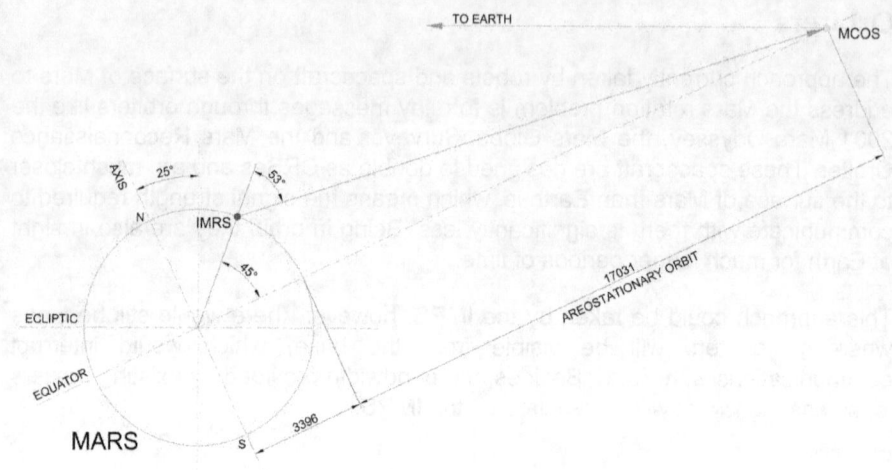

MCOS in areostationary orbit above the IMRS

Streaming video from Mars

One of the advantages of an areostationary orbit is that MCOS can provide, not only communications, but also continuous observation of IMRS.

MCOS will include a high resolution colour camera comparable or superior to HiRISE (High Resolution Imaging Science Experiment), enabling the base to be monitored from orbit. This will be especially important in case of emergency. If other modes of communications break down, being able to see the IMRS from space will provide valuable clues as to the situation on the surface (a scenario described in Andy Weir's superb novel *The Martian*).

Such a camera should be both cheaper and technologically superior to the $40 million HiRISE camera, due to evolution in CCD (Charged Coupled Device) and other related technology.

MCOS will ideally also provide real-time HD (High Definition) video from space. The live HD video stream of Earth from the ISS provided by UrtheCast will be instructive in the design of such a system.

With at least two cameras available, MCOS will be usable for more than simply observation of the base. It can be used for studying almost one entire side of Mars and provide advanced notice of weather conditions, such as dust storms, that could affect a mission.

A crucial benefit is that live images and video beamed direct from IMRS, from cameras on the ground as well as from MCOS, will greatly increase engagement in the mission. This, in turn, increases ROI, since the higher the level of engagement, the more people are benefiting in terms of inspiration, education, and contributing their attention to the evolution and success of humanity.

The view from MCOS won't be straight down onto the base because it will be above the equator, whereas IMRS is more likely to be located between about 30°N and 60°N, as discussed in Site Selection. As shown in the above drawing, if IMRS is located at latitude of 45°N, the view from MCOS will be at an angle of about 53°, which is still a perfectly useful viewing angle.

It's worth noting that the potential for an areostationary communications satellite is yet another important advantage of the "base-first" strategy, as this would not be practical if each mission was to a completely different location. However, with a plan to build up a base at a single location, it makes perfect sense. Having said that, with continuous visibility of almost 42% of the planet the MCOS could potentially serve multiple missions or bases on the same side of Mars.

MARSNET

The addition of a second MCOS in aerostationary orbit at a different longitude would provide connectivity between the IMRS and Earth during those occasional brief periods when Mars blocks the line of sight between MCOS and Earth.

At those times, communications can be relayed to "MCOS-2", and from there to Earth. In fact, by placing MCOS-2 at 120° from MCOS-1, 100% connectivity can be provided between Earth and about two-thirds of the Martian surface. Each MCOS would function as a relay for the other when required.

If there was only to be one base on Mars then it would be hard to justify the expense of a second MCOS. However, because MCOS-2 has the potential to open up another one third of the planet, in addition to enabling 100% connectivity with IMRS, it should really be added as soon as budget permits.

In fact, near global coverage of Mars can be provided by three such satellites, separated from each other by 120°. This will provide communications connectivity with Earth from almost everywhere on Mars except for the very highest latitudes (i.e. close to the poles).

Such a setup would form the early stage of a global Martian internet, or MARSNET. There's little doubt that Martian internet will be satellite-based. Earth will have global satellite-based internet within about 5-10 years, which will drive the evolution of space-based communications.

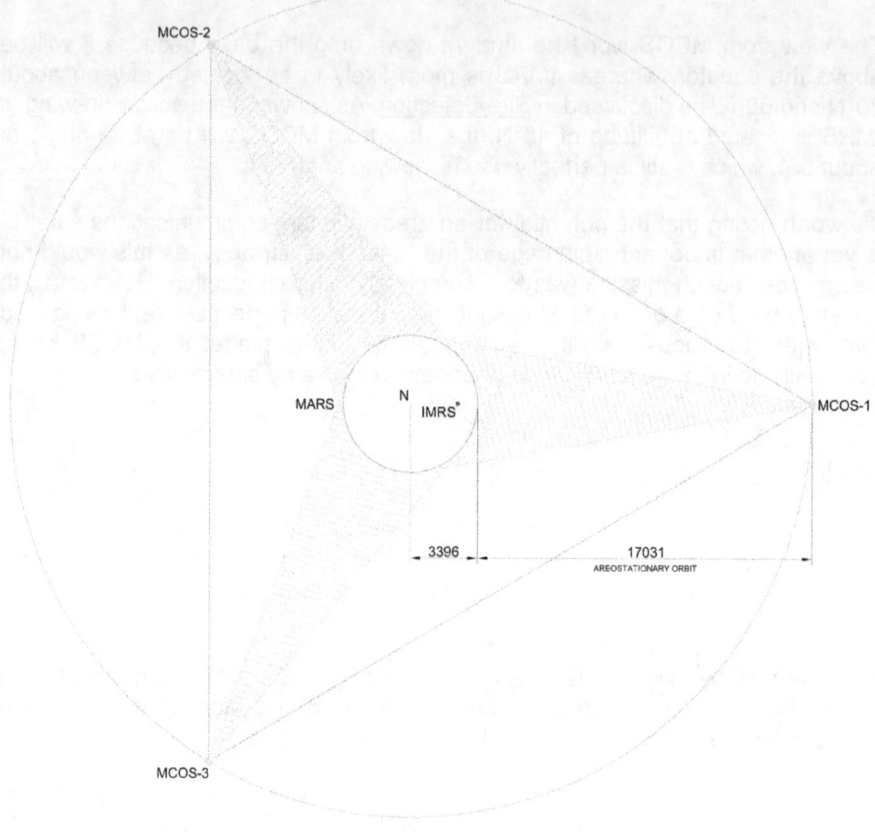

MARSNET 0.1

The construction of this simple satellite network will be invaluable for ongoing exploration and settlement across the planet, capable of supporting both robotic and human missions for many years.

One potential schedule would be to deploy one MCOS with each of the first three missions. This would enable 99% connectivity with Earth from the first mission, and 100% from the second mission. From the third mission, almost all of Mars would be accessible to humans in terms of continuous communications with Earth. However, this timeline could be unnecessarily rapid in terms of budget, and the third satellite could be deployed later, when specifically required by future missions.

Mars One solution

The solution developed for Mars One (Lansdorp & Wielders 2014) to provide approximately 99% connectivity between Earth and the Mars One settlement relies on one CRS for each of the problems described earlier.

To address the problem of Mars's rotation, the same idea of using an areostationary satellite above the base is planned.

To address the problem of solar conjunction, a satellite is to be placed at Earth's L5 Lagrangian point (i.e. sharing Earth's orbit, but trailing by 60°), which will relay communications around the Sun during this period.

The advantage of placing a CRS at L5 is that this is a gravitationally stable Lagrange point, which reduces or eliminates the need for station-keeping. However, this apparent benefit is a double-edged sword, because L4 and L5, being gravitationally stable, tend to attract meteoroids and micrometeoroids (space dust). At L4, which is 60° ahead of Earth in the same orbit, is the asteroid 2010 TK$_7$, making it an Earth trojan. There are no known Earth trojans at L5, however, rocks and dust are likely, and these could potentially damage the satellite.

MARSLINK

The solar conjunction problem could perhaps be solved more effectively with a set of three HLCRSes, separated from each other by 120°, and thus forming a rotating equilateral triangle, similar to the three MCOSes around Mars. These three satellites may be referred to as MARSLINK.

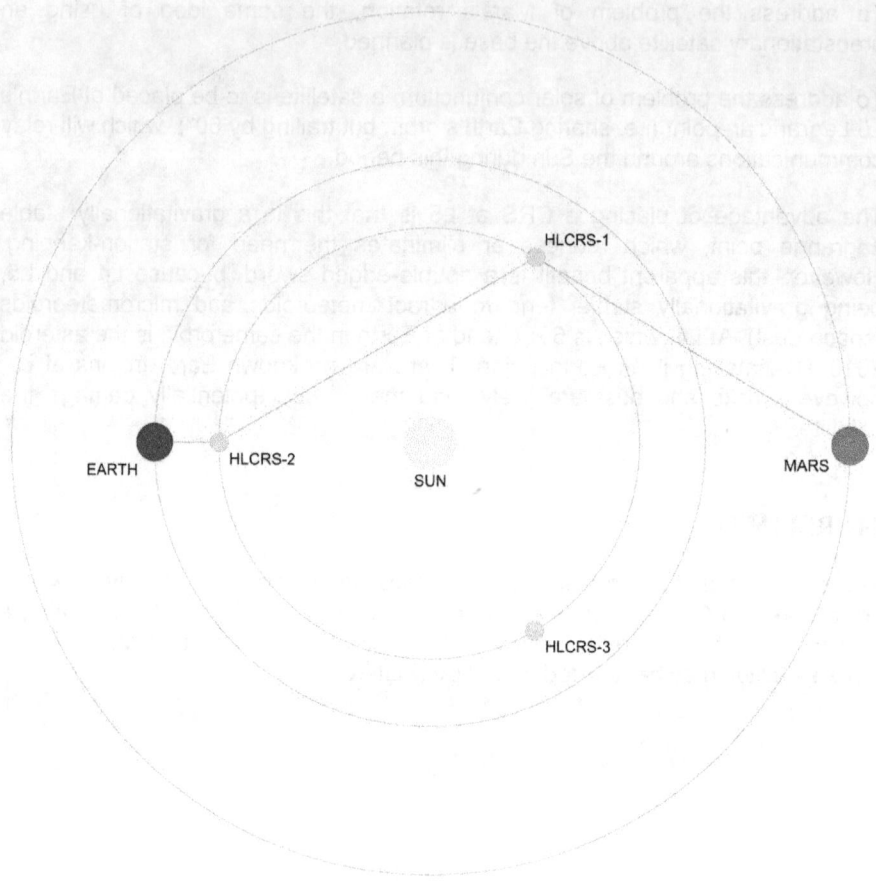

MARSLINK 0.1

Whenever Earth and Mars are separated by more than, say, 90°, communications between the planets can be routed through one or two of these satellites. The following diagram illustrates some possible routing paths given different relative positions of Earth, Mars and the satellites.

Possible routing pathways between Earth and Mars

This solution to the solar conjunction problem is more expensive than the single L5 satellite, requiring three satellites instead of one. However, the HLCRSes may not be too expensive, and there are numerous advantages to this approach:

1. Less likelihood of being damaged by meteoroids.

2. Maintaining a constant orbital radius is less important in this configuration, as is the satellites' velocity. Even if a satellite's station-keeping mechanism malfunctions and it begins changing altitude, it will not move quickly away from its original orbit (which is quite stable) and will continue to function perfectly well for many years. The satellites do not need to remain in a fixed position with respect to Earth or Mars, but only to be visible.

3. Redundancy is built in. If one of the satellites goes offline for any reason, or is blocked by another object, then there is always an alternate route;

either directly to the target (Earth or Mars), or the other way around the triangle.

4. The satellites will be solar-powered. Being a little closer in to the Sun than Earth orbit will be advantageous for boosting signal strength.

5. An unlimited number of additional satellites can be added to either the same orbit or others (including heliocentric polar orbits), in order to improve redundancy, reduce lag times, increase bandwidth, decrease transmission distances, or improve signal strength.

6. The satellites will be well-positioned to connect, not only Earth and Mars, but all inner Solar System destinations, including the Moon, Venus, Mercury, NEAs, Ceres, etc.

7. If so designed, the satellites could fulfil additional functions, such as solar observation or asteroid hunting. They could, in fact, serve as an early-warning system for CME's, relaying information directly back to Mars to warn the crew in advance of impending solar events.

Orbital parameters

In the above drawings of MARSLINK, the HLCRSes are in a circular orbit at approximately 125 Gm, roughly halfway between the orbits of Venus (103 Gm) and Earth (150 Gm). This radius is optimal for minimising the longest transmission distance between Mars and the satellites. The benefit of minimising these transmission distances is that transmitters and receivers will not need to be as powerful, making them cheaper, lighter, and less power-hungry; or, alternately, the bandwidth can be greater.

However, there are other factors that could contribute to determining the ideal orbital radius. For example, Earth, with its large radio antennas, will be able to hear much fainter signals than the satellites; therefore, most of the time it will be more efficient, in terms of minimising communications lag time, for packets to be transmitted directly between Earth and Mars when practical (as shown in the lower 4 images, above).

Another factor is power. The satellites will be solar-powered, therefore, they closer they are to the Sun the more energy they will be able to harness for relaying transmissions and perhaps also for station-keeping. The transmission line-of-sight does not have to miss the Sun by much to avoid thermal noise, suggesting that the satellites could be quite close in, perhaps even inside the orbit of Venus.

The downside of being too close to the Sun is that the satellites would be exposed to a higher intensity of solar radiation and could degrade faster; also, they'll be more greatly affected by solar radiation pressure, which will increase the amount of propellant required for station-keeping. Both of these effects would reduce the satellites' lifetime.

In addition, the closer they are to the Sun, the faster they'll be moving. This could increase the difficulty of targeting a steerable communications laser with accuracy. If the satellites are farther out from the Sun, they will be closer to Mars, and moving slower, which should, in theory, improve the accuracy of transmission.

Further analysis is necessary to determine the optimal orbital radius for these satellites.

MarsSat

The main problem with MARSLINK will be the mass and power requirements of the satellites, considering the large distances between them and Mars. Not only would the satellites in heliocentric orbit need to be reasonably powerful, and thus heavier and more expensive, but also the ones in Mars orbit.

A cleverer solution for maintaining Earth-Mars communications during solar conjunction could be "MarsSat" (Gangale 2006). MarsSat is a CRS occupying a heliocentric orbit with the same period as Mars, except with a slightly different eccentricity and orbital plane. This type of orbit is called a "Gangale orbit", after its designer, Tom Gangale. Such a satellite would appear, from the perspective of Mars, to orbit Mars once per Martian year, despite being gravitationally bound to the Sun.

The benefits of MarsSat:

- As it is not located at Earth or Mars L4 or L5 points, it would not be susceptible to damage from trojan meteoroids or micrometeoroids (although naturally it would still be at risk of collision with any of the countless small objects circulating through the inner Solar System, as are all satellites and spacecraft).

- The distance between Mars and MarsSat would be on the order of 22 Gm. As this is around one tenth the maximum distance between Mars and the closest MARSLINK satellite, or one at L4/L5, the signal received by MarsSat would be around 100 times stronger. This means both the MCOS transmitting antenna and the MarsSat receiving antenna can be smaller, which translates to lower satellite masses and power requirements, and therefore lower cost.

Analysis has shown (Byford, Goppert & Gangale 2008) that the optimal configuration for an Earth-Mars communications link, in terms of mass, power and cost, is three CRSes in areostationary Mars orbits (like MARSNET) plus two additional satellites in Gangale orbits, leading and trailing Mars by 8° respectively. All communications with Earth would be routed through the satellites in the Gangale orbits, significantly reducing the power requirements and therefore the masses of the satellites in orbit around Mars. As the cost of placing satellites in Mars orbit is higher than the cost of placing them in Gangale orbits, this approach greatly reduces the overall cost of the system.

11.3.2. Between Earth and *Adeona*

Adeona will be equipped with deep space communications equipment. During both the outbound and return journeys, Earth and Mars will be on the same side of the Sun, and therefore line-of-sight radio communications between Earth and *Adeona*, Mars and *Adeona*, and Earth and Mars, will be possible at all times during the space element.

When *Adeona* is in Earth orbit below about 3000 km altitude, communications between the spacecraft and Earth can be provided by the SN (Space Network), which provides 100% coverage up to this altitude. The SN is comprised of the geosynchronous TDRSS (Tracking and Data Relay Satellite System) in addition to various ground-based stations.

When *Adeona* is in Earth orbit above about 30,000 km altitude, the DSN (Deep Space Network), comprised of large radio antennas in California, Spain and Australia, can provide a continuous communications link between the spacecraft and Earth.

There is currently no space communications network providing full coverage between 3000 and 30,000 km altitude. However, *Adeona* should only be between these altitudes for short periods, as it climbs out of LEO prior to TMI, and as it returns to LEO after EOI. During this time, when neither SN nor DSN can provide 100% coverage by themselves, it may be that the burns are timed such that communications can be provided by either the SN or DSN or both working together, or that communications are provided by other space communications systems such as the Russian *Luch* satellites, or that the crew and MCC can accept reduced comms for these brief periods.

Once *Adeona* is in Mars orbit, it can make use of MARSNET and MARSLINK to relay telemetry data back to Earth.

11.3.3. Surface Communications

Between SHAB and CAMPER

The SHAB and CAMPER will have line-of-sight radio comms when exploring near the base. When the CAMPER is out of direct radio contact with the SHAB — for example, when it is over the horizon or behind a hill — communications can be relayed via MCOS.

Both the SHAB and CAMPER will have a UHF antenna with software-defined radio for communication with the MCOS and with each other, possibly in addition to an X-band antenna for direct communications with Earth.

As part of ongoing development of IMRS, one or more mobile towers may be installed around the base, which could function as an LPS as well as communications relays, functions that will both be especially useful during long-range excursions in the CAMPER.

Between astronauts and SHAB/CAMPER during EVA

Astronauts on EVA will always maintain radio contact with either the CAMPER or SHAB or, ideally, both. In the rare case that this rule needs to be broken — for example, while exploring a lava tube or cave — then the pair or group (assuming that the crew always work in groups of at least two) must remain within radio contact of each other at all times, with at least one of them in radio contact with the SHAB or CAMPER.

11.3.4. Internet protocols

Internet technology will be used for most of the communications requirements throughout the mission.

Inside the THAB, SHAB and CAMPER, and around the base, standard TCP/IP (Transmission Control Protocol/Internet Protocol) — the protocols used for internet services on Earth — and ordinary wifi technology will be practical.

Long-range communications, however, will not use TCP/IP. Communications between Earth, satellites, *Adeona* and the IMRS will use BP (Bundle Protocol), an internet protocol specifically developed for DTN (Delay/Disruption Tolerant Networking). BP is designed for situations where delays and disruptions to communications are expected, such as space missions. One of the functions of the ISS has been to develop and improve this technology.

The Interplanetary Internet

MARSNET and MARSLINK could potentially form part of the Interplanetary Internet, a plan for a space-based communications network that will link the worlds of the Solar System. It will be a network of CRSes, probably based on laser communications and BP, located in strategic orbits throughout the Solar System and providing internet connectivity between Earth and all spacecraft and settlements in the System.

11.4. Marssuits

11.4.1. Mechanical Counter-Pressure Suits

Blue Dragon does not incorporate traditional gas-pressurised spacesuits. Instead, it's assumed that MCP suits, also known as SASes (Space Activity Suit), will be available in the timeframe of the mission, as these are an active field of research and considerable progress is being made.

MCP spacesuits are skintight, somewhat like SCUBA (Self-Contained Underwater Breathing Apparatus) wetsuits, with a solid helmet and backpack. The backpack, helmet and hard-shell parts of the suit contain life-support, communications, navigation and computing technology.

Perhaps the most advanced MCP suits currently in development are the BioSuits being researched at MIT (Massachusetts Institute of Technology) under the leadership of Dava Newman, Professor of Aeronautics and Astronautics and Engineering Systems at MIT, pictured below.

Full BioSuit (Credit: MIT)

MCP suits offer important advantages over traditional gas-pressurised spacesuits:

1. They provide considerably better mobility. Gas-pressurised suits are essentially a rigid balloon, and bending any of the joints requires constant force. The gloves are particularly hard to use, comparable to squeezing a tennis ball. Use of the suits is therefore rather exhausting, and the astronauts cannot work in them for long periods. MCP suits, in comparison, are much less bulky, more flexible, and easier to move around in.

2. They are significantly lighter. The Apollo suits weighed 91 kg, and the new

NASA Z-2 gas-pressurised spacesuits weigh 65 kg; but a fully charged BioSuit including helmet and PLSS is expected to weigh only 20 kg. On Mars, this will feel like just 7.6 kg, which is well within comfortable limits (an adult backpacker will typically carry at least 2-3 times this load). This low mass means EVAs can be longer due to reduced fatigue, and are safer due to lower inertia and therefore reduced risk of falling over. It will permit astronauts to lift and carry equipment and samples while on EVA. It also means the suits are cheaper to transport, and complete spares may be taken.

3. They can provide a higher air pressure. Gas-pressurised suits typically provide very low O_2 pressure in order to maximise mobility as much as possible. However, it's unhealthy to be exposed to low O_2 pressure for extended periods, which limits EVA duration. In addition, it means the astronauts need to spend lengthy periods, sometimes hours, pre-breathing pure O_2 to purge N_2 from their blood so that they don't get DCS. BioSuits will provide about 35 kPa of pressure; any higher and mobility begins to be compromised. This level of pressure is much more healthful for the astronauts. Significantly, with an appropriate habitat atmosphere design, it is possible to use the suits without prebreathing, as discussed in SHAB Atmosphere.

Overall, the use MCP suits means that astronauts will be able to work outside more frequently and for longer periods, more healthfully and comfortably, and while capable of doing more useful work. These suits are an absolute necessity for Mars missions, which are long duration with an emphasis on outside work. Literally hundreds of EVAs may be required per astronaut.

BioSuit compared with traditional gas-pressurised suit (Credit: MIT)

Replacement parts and patch kits will be included in surface mission supplies, and the SHAB's server will be preloaded with models of all spare parts that can be printed by a 3D printer.

11.4.2. Suit Pressure

From private communications with Prof. Dava Newman:

> *We've made some outstanding progress this semester on our active materials and believe we can now achieve the 30 kPa pressure level. It shouldn't be asking too much for us to achieve the mid-30 kPa, especially by 2030!*

A useful target pressure for the marssuits is approximately 35 kPa. As discussed in SHAB Atmosphere, this is the suit pressure needed for interoperation with the proposed habitat atmosphere, in order to establish a ZPB protocol and eliminate risk of DCS, irrespective of the number of EVAs.

According to Prof. Newman, it's unlikely that suit pressure much higher than 35 kPa will be sought as this would begin to compromise mobility, which is the primary driver of MCP suit design. Fortunately, this is enough.

Oxygen toxicity

A suit pressure of 35 kPa is within safe limits for O_2 toxicity. It has been reported (Campbell 1991) that O_2 toxicity can occur when exposed to a high O_2 partial pressure for an extended period; specifically, above 40.5 kPa for more than approximately 24 hours. The same report also states:

> For EVAs of six to eight hours, 100 percent oxygen can be used at pressures up to 41.4 kPa (6.0 psia) with absolutely no risk.

As the proposed suit pressure falls below this threshold, oxygen toxicity is not a risk. This highlights another advantage of the MCP suit over gas-pressurised ZPB suit designs such as the new Z-2 spacesuit, which has an operating pressure of 57.2 kPa. The risk of O_2 toxicity makes marssuits with O_2 pressure higher than about 40 kPa less desirable for HMMs.

11.4.3. Suit Storage and Recharging

Inside both the SHAB and CAMPER will be a suit storage and recharge bay with capacity for up to six marssuits.

The suits will be powered by rechargeable graphene batteries or supercapacitors. In the suit recharge bay in the SHAB, there will be charging sockets where the suits' battery packs can be recharged from the SHAB's power system. In addition, hoses in the recharge bay connect the suits' O_2 tanks to the SHAB's tanks for refilling.

Valves will be made available on the outside of the SHAB to enable refilling of marssuits tanks from the SHAB's supplies of breathing gases, without the need to re-enter the habitat. This could be important in an emergency, such as an astronaut running very low on O_2. In addition, the external valves will enable refilling of the CAMPER's breathing gas tanks between excursions, by parking alongside the SHAB and connecting a hose.

11.5. Airlocks

Both the CAMPER and SHAB will have an airlock capable of accommodating up to two astronauts at a time. Most EVAs will involve two people, so this is a reasonable airlock size considering the need to optimise for mass, volume and time.

The B330 modules include two airlocks; one at each end.

11.5.1. Minimising Dust Migration

A feature of modern gas-pressurised spacesuit design such as the NASA Z-2 suit is the suitport, which enables spacesuits to remain outside. The suitport connects to a hatch on the side of a rover of habitat, and astronauts don the suit by climbing through the hole and into the suit. In this way the suit always remains outside and airlock operation can be avoided.

Suitports on the rear of the Small Pressurized Rover (Credit: NASA)

One of the primary advantages of suitports is that they prevent dust migration into the vehicle or habitat. Lunar dust that found its way into the Apollo Lunar Excursion Modules caused the astronauts to experience a number of respiratory issues, plus the dust was highly adhesive and hard to remove from suits and other surfaces.

Martian dust is different from lunar dust. Although also very fine, it is far less adhesive due to being less jagged, as erosion caused by the continual winds smooths the particles. However, it can still cause problems.

Perhaps the most serious risk associated with Martian dust is its high concentration of perchlorate, the ClO_4^- ion. Perchlorate inhibits thyroid function, and is therefore sometimes prescribed as a treatment for hyperthyroidism. If ingested, it can interfere with iodine uptake, hormone production and regulation of metabolism.

Martian dust may additionally cause respiratory issues; interfere with air quality, experiments, electronics, ECLSS or other SHAB systems; generally make things dirty; or cause other, unforeseen problems. Dust migration into the SHAB and CAMPER must therefore be kept to a bare minimum.

In order to mitigate dust migration into the SHAB and CAMPER, their airlocks double as suit-cleaning areas prior to re-entering.

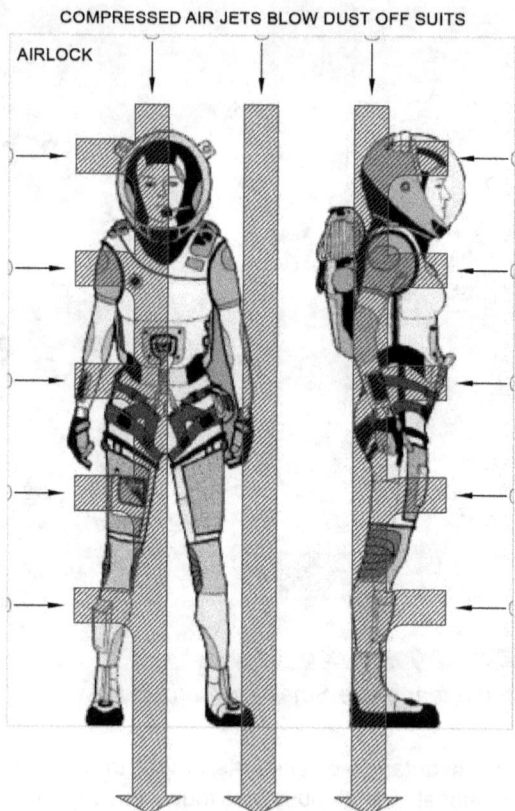

Air flow and dust removal in airlock

11.5.2. Airlock Operation

The airlock has three modes of operation, controlled by buttons aptly coloured red, green and blue:

Red: For exit. Pump breathable air out of the airlock.

Green: For entry. Dust cleaning process.

Blue: For entry. Fill the airlock with breathable air.

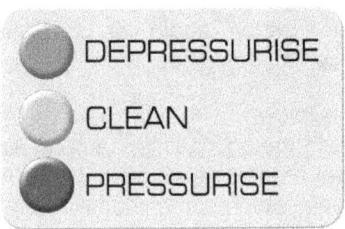

Airlock controls

A wall-mounted screen shows the current pressure in the airlock, and concentrations of O_2, CO_2, N_2 and Ar. The display indicates "SAFE" when the air in the airlock can safely be breathed. This is a precaution only, as during normal operation astronauts will breathe from their suits when inside the airlock.

Indicators on the inner and outer doors display "PRESSURE MATCH" when the airlock pressure matches either the SHAB interior pressure (~53 kPa), or the external pressure (~0.6 kPa), respectively.

The SHAB's computer system controls airlock operation, as with all systems in the module.

Exit process

1. Two crew members take their marssuits from the suit recharge bay, don them, seal the helmets and commence breathing pure O_2 from the suits.

2. They step into the airlock and close the inner door.

3. They press the red button. If the computer cannot verify that both inner and outer door are closed, a warning is displayed. Once both doors are secure, the air is pumped out of the airlock, down through the floor vent.

4. The pressure is reduced until it matches the external pressure, at which point the pressure-match indicator on the outer door will light up, and the computer will permit the outer door to be opened (this can be manually overridden in an emergency).

5. The astronauts open the outer door and step out onto Mars.

Entry process

1. On returning from EVA, the astronauts kick the dust off their boots and/or wipe their feet on a mat outside the entrance to the SHAB airlock.

2. They open the airlock outer door, enter the airlock, and close the door.

3. They press the green button. Again, the computer verifies that both doors are secure before commencing operation.

4. Compressed gas is blown in through jets in the ceiling and walls, and sucked down through the floor vents, carrying away the dust and blowing it outside. This is not breathable air from the SHAB, which costs energy to make, but Martian air, which is more easily obtained. When the airlock is not in use, air is pumped in from outside, filtered, compressed and stored in a tank adjacent to the airlock for this purpose.

5. The compressed air jets are supplemented by two powerful handheld vacuum cleaners with brushes, which the two astronauts use to remove any remaining dust from their own and each other's suits. They work together to clean the suits and the airlock interior as meticulously as possible until both are fully satisfied. Despite the flow of air, the pressure does not substantially change during the cleaning process, remaining at approximately ambient Martian pressure.

6. They then press the blue button. The floor vents close, and the gas from the compressed air jets switches from Martian air to breathable air from the SHAB's tanks.

7. When the pressure in the airlock matches the interior of the SHAB, the jets shut off, the pressure-match indicator on the inner door lights up, and the computer permits opening of the inner door (again, this can be manually overridden in an emergency).

8. The two astronauts enter the SHAB, doff their clean marssuits, and store them in the suit recharge bay where their batteries are recharged and O_2 tanks refilled.

The airlock's dust filter is cleaned manually or automatically at the end of each EVA, or periodically as required.

12. Precursor Missions

Apollo missions 11-17 succeeded because of Apollo missions 4-10 and all the other missions that preceded them. The research, hardware tests, flights and missions that preceded the human landings tested each element of the target mission, progressively improving confidence in the architecture, hardware, systems, processes and astronauts.

Precursor missions are not often described in conjunction with HMM architectures, perhaps because of the exorbitant cost of sending packages to Mars, or perhaps because the human mission itself is complex enough to describe. Perhaps it's because precursor missions are considered less of a challenge or already addressed by robotic exploration.

However, once an architecture has been selected, it will be necessary to test as much of it as possible before attempting a human landing, just as with Apollo.

If the Red Dragon technology can reliably deliver 2-tonne payloads to Mars surface for only a few hundred million dollars per launch, a program of precursor missions that prove and develop the technology, in addition to testing other elements of the human mission, is affordable and can be included in the overall program.

A schedule of precursor Mars missions will enable improvement of the architecture. Many aspects of Blue Dragon can be tested on Earth, in Earth orbit, or on the Moon, but some are only truly testable at Mars, especially those related to EDL and ISRU. Examples include the following, some of which can be bundled together into the same mission:

- Test of the MAV, including EDL, ISPP, launch and ascent to orbit (everything up to, but excluding, MOR).

- Test of an ISAP system to produce breathable air from Martian atmosphere.

- Test of ISWP systems to extract water from the Martian atmosphere and ground.

- Test of solar power production on Mars using PV panels, material and coatings.

These are combined into the following proposed missions:

Green Dragon: a robotic lander that tests entry, descent and precision landing of a Red Dragon capsule, and all ISRU systems required by the SHAB, i.e. in situ production of electricity, air and water. Water and gases are obtained from the atmosphere and processed to produce breathable air and potable water. Power production using PV panels, blankets and/or coatings, and potentially also ASRGs, is also tested. If mass, volume and budget constraints permit, the mission may additionally test production of CH_4 and/or CO, and the use of these in fuel cells to produce electricity, which in turn can provide power to the lander.

Gold Dragon: a MAV test. A MAV prototype is landed on Mars and the ISPP unit is activated, producing LOX/LCH4 bipropellant. Once full, the MAV is launched from the surface of Mars, and ascends to an orbit matching where *Adeona* will be, as if in preparation for MOR. This mission will test EDL of the MAV, PV blankets and coatings, ASRGs, the AWESOM robot, ISPP, and autonomous launch and ascent of the MAV.

These missions can benefit from the research already conducted for the MARCO POLO, Ice Dragon, Mars 2020 rover, and other related missions.

In a perfect Universe, a complete run-through of the mission with no crew involved would be a valuable test. After a successful predeployment phase, *Adeona* could be constructed and robotically flown out to Mars where it would wait on orbit for 1.5 years, then the MAV would be launched, the ascent capsule docked with the MTV then undocked, and the MTV robotically flown back to Earth and parked on Earth orbit. All of this could actually be done without a crew. However, this would be a very lengthy and expensive test, and unlikely to be practical or truly necessary. A combination of simpler precursor missions, simulations, and Earth-based missions will suffice, as they did with Apollo.

12.1. Green Dragon

Green Dragon is a precursor mission to the HMMs, which will land a Dragon capsule on Mars containing integrated ISRU hardware capable of manufacturing breathable air and potable water from local Martian resources. The mission is designed to advance capabilities and build confidence in two main areas: EDL and ISRU.

12.1.1. Entry, Descent and Landing

One of the main reasons for Green Dragon is to gain practice landing Red Dragon capsules on Mars, refine the technology, and build confidence that this can be done repeatably and safely. If Mars One is successfully implemented before the IMRS program is commenced, then EDL via capsule may already have been proven. In any case, the more this is practiced, the better. Ideally, at least a few capsules will have been successfully landed on Mars before using one for landing a crew:

- 1 for Green Dragon
- 2 or more supply capsules

- Any capsules sent by Mars One

The more capsules landed on Mars before sending humans, the more confidence mission planners will have in the technology before landing a crew using the same technology. It's necessary to know exactly how capsules behave during EDL to Mars, how they interact with the Martian atmosphere and surface, how to land on a dime, and where the strengths and weaknesses are in this approach. Although much can be simulated, nothing will deliver as much insight as actually landing a capsule on Mars.

It's essential to test and, where possible, improve the accuracy of landing. SpaceX claim that Dragon capsules will be capable of pinpoint landings on Mars; other navigation algorithms that could be tested in this mission include LION and G-FOLD.

12.1.2. Integrated ISRU

The other main purpose of the Green Dragon program is to increase the TRL of ISRU technology in preparation for sending humans to Mars. The intention is to test ISRU systems for producing breathable air (O_2 and buffer gas), potable water, and potentially also CO for use in a fuel cell. A secondary goal is to create a backup cache of critical resources near the IMRS that may be accessed in case of emergency. Green Dragon effectively simulates ISRU systems for the SHAB.

Green Dragon includes three integrated ISRU experiments:

1. **ISEP (In Situ Electricity Production)** - Electrical power is produced from solar energy using one or more rolls of PV material.

2. **ISAP** - O_2 and CO are obtained by electrolysis of atmospheric CO_2 using SOECs. Buffer gas comprised primarily of N_2 and Ar is produced by drying and detoxifying the gas mix that remains after CO_2 is removed from Martian air. The CO is retained for use in a fuel cell, producing additional electricity to power the spacecraft.

3. **ISWP** - Water is extracted from the atmosphere via adsorption in zeolite 3A.

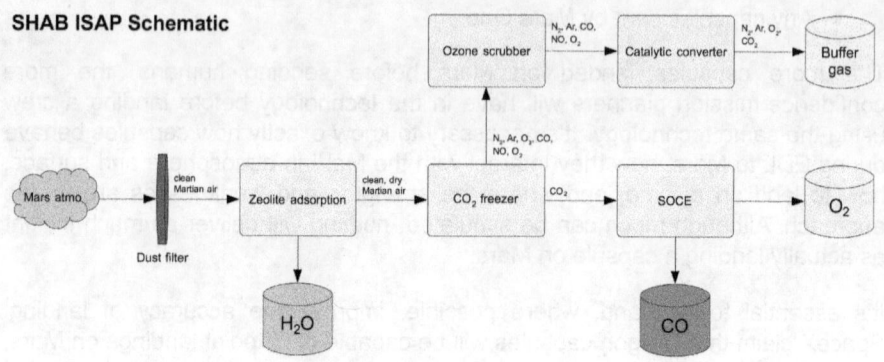

Green Dragon integrated ISRU schematic (same as for the SHAB)

12.2. Gold Dragon

Gold Dragon is simply an automated MAV test.

Launching from the surface of Mars is one of the greatest challenges in HMMs, which is why it was omitted from Mars One, and why some people favour one-way human missions.

However, considering SpaceX's RLS (Reusable Launch System) technology, Masten Space Systems' Xombie rocket, Blue Origin's New Shepard VTOL rocket, and other developments currently underway in space, energy, materials and manufacturing, development of a MAV in the required time frame definitely seems achievable.

Operation of the MAV is more-or-less impossible to properly test on Earth, and it would be risky (perhaps unacceptably so) to attempt to launch humans from the surface of Mars without having tested this launch system at least once.

The Gold Dragon MAV will be flown to Mars, landed somewhere close to the planned IMRS location, filled with propellant over a period of months, and launched to Mars orbit. As there will be no MTV on Mars orbit at the time, the only part of the MAV lifecycle that will not be tested by this mission will be the docking of the Mars Ascent Capsule (i.e. *Kepler*) with *Adeona*. After being placed in the 250 km x 1 sol orbit, it may be left there for later recovery, or de-orbited to fall back to Mars.

The reason why the Gold Dragon MAV needs to be landed close to the IMRS is to test water extraction from the same latitude and type of terrain that the MAVs used in the human missions will encounter.

The intention is to send a Gold Dragon mission at least once, and perhaps again if results indicate that the vehicle needs to be further improved before it can be declared human-rated. While this would be expensive, the ISPP system and

MAV ascent are firmly in the critical path of the mission and it's essential they work as planned.

Abbreviations

Chemical Elements

Chemical symbol	English name	Notes
Ar	argon	A noble gas, non-reactive, but toxic in high concentrations. About 1% of Earth's atmosphere and 2% of Mars's. Used in welding.
C	carbon	One of the most useful and abundant of all elements. Definitive element of organic molecules. Found in wood, plastic, steel, diamond, nanotubes and many other materials.
H	hydrogen	The lightest and most abundant element in the Universe. The main element in stars, commonly found in water and hydrocarbons.
N	nitrogen	An essential element for life, found in all living things. Comprises about 80% of Earth's atmosphere.
O	oxygen	An essential element for life, and extremely abundant. Most often found in water, air, and in rocks as metal oxides.
Pu	plutonium	A radioactive metal sometimes used as power source in mobile robots and spacecraft.

Chemical Compounds

Chemical formula	English name	Notes
CH_4	methane	Simplest hydrocarbon, non-toxic, major constituent of natural gas, great fuel.
CH_3OH	methanol	The simplest alcohol. A light, volatile, colourless, flammable liquid that can be used as a fuel or to make other chemicals.
CO	carbon monoxide	Toxic gas found in car exhausts. Can be used in fuel cells and iron refining.
CO_2	carbon dioxide	Non-toxic in small concentrations, exhaled by animals and consumed by plants (photosynthesis). Comprises about 96% of the atmospheres of both Mars and Venus.
H_2	hydrogen	The most abundant and lightest gas in the Universe.
H_2O	water	Essential for life, extremely abundant in the Universe, an excellent solvent, we're mostly made of it, Earth is mostly covered in it, you can swim in it, wash with it, throw it on people during Thai New Year, even drink it.
N_2	nitrogen	Non-toxic and a good buffer gas. Comprises about 78% of Earth's atmosphere, 2% of Mars's, 3.5% of Venus's, and 95% of Titan's.
NH_3	ammonia	A metabolism product and important component of fertiliser.
NO	nitric oxide	Important biological regulator relevant to neurology, physiology and immunology. Non-toxic, but rapidly oxidises to toxic NO_2 in the atmosphere.
NO_2	nitrogen dioxide	A brown toxic gas and major air pollutant, commonly produced by internal combustion engines.
O_2	oxygen	Inhaled by animals, exhaled by plants. Extremely abundant in the Universe, comprising about 20% of Earth's atmosphere. Most common oxidiser used in rocket propellant.
O_3	ozone	A toxic form of oxygen that absorbs ultra-violet radiation. Found in the upper atmosphere.

Acronyms and Initialisms

Numerous acronyms and initialisms appear in this document; all are listed below for easy reference. They are drawn from several domains, including space, technology, business and military. Several have been newly invented for this document.

To mitigate acronym overload, and reduce time spent flicking back and forth to this page, acronyms are expanded on first usage.

AEB	Agência Espacial Brasileira	Brazilian Space Agency.
AG	Artificial Gravity	Sensation of gravity, usually produced by centripetal force.
ASRG	Advanced Stirling Radioisotope Generator	Stirling engine powered by a large radioisotope heater unit.
ATV	All Terrain Vehicle	Quad bike suited for driving on rough terrain, such as on Mars.
AWESOM	Autonomous Water Extraction from the Surface Of Mars	Method proposed for extracting water from the Martian regolith using a mobile robot.
BP	Bundle Protocol	Internet protocol designed for DTNs.
CAMPER	Crewed Adaptable Multipurpose Pressurised Exploration Rover	A pressurised rover for multi-sol excursions on Mars, featuring robotic arms and attachments such as a drill and excavator bucket.
CCD	Charge-Coupled Device	Component used in digital cameras for detecting light.
CEO	Chief Executive Officer	Senior executive position in a company.
CIGS	Copper Indium Gallium Selenide	Semi-conductor material used as a substrate on flexible, thin-film solar cells.
CME	Coronal Mass Ejection	Massive burst of electrons, ions and atoms from the Sun's corona into space. Also known as a solar flare.
CNSA	China National Space Administration	Space agency of China.
COSPAR	Committee for Space Research	Committee that promotes international collaboration with respect to scientific research in space.
COTS	Commercial Off The Shelf	Component that can be purchased from a commercial vendor, rather than custom-designed and built from scratch.
CRISM	Compact Reconnaissance Imaging Spectrometer for Mars	A visible-infrared spectrometer aboard the Mars Reconnaissance Orbiter searching for mineralogic indications of past and present water on Mars.
CRS	Communications Relay Satellite	A satellite that receives and retransmits communication signals.
CSA	Canadian Space Agency	Space agency of Canada.

DCS	Decompression Sickness	Illness caused by dissolved gases (usually nitrogen) forming bubbles in the blood upon depressurisation.
DRA	Design Reference Architecture	NASA's fundamental architecture and assumptions for sending humans to Mars.
DSN	Deep Space Network	Worldwide network of large antennas and communication facilities located in California, Spain and Australia, used for tracking interplanetary spacecraft.
DTN	Delay or Disruption Tolerant Networking	Approach to computer network architecture where the network may be subject to delays or disruption, such as in space.
ECLSS	Environment Control and Life Support System	System for maintaining a vehicle or habitat's environment at the right conditions to support human life.
EDC	Earth Descent Capsule	Crew Dragon launched from Earth at the end of the mission, which the crew use to land on Earth. Also known as *Newton*.
EDL	Entry Descent and Landing	Process of landing a spacecraft on a planet with an atmosphere.
EDRS	European Data Relay System	Also known as "SpaceDataHighway". Collection of satellites in GEO providing high bandwidth laser communications between satellites, spacecraft and ground stations.
EELV	Evolved Expendable Launch Vehicle	Expendable launch system program designed to assure access to space for US government launches, and make them more affordable and reliable.
EMC	Earth-Mars Capsule	Crew Dragon that carries the crew from Earth surface to Earth orbit, then from Mars orbit to Mars surface. Also known as *Einstein*.
EOI	Earth Orbit Insertion	Orbital manoeuvre that moves a spacecraft into Earth orbit.
EOR	Earth Orbit Rendezvous	When two spacecraft meet and dock in Earth orbit.
EP	Electric Propulsion	Propulsion method that uses electricity to produce positive ions, which are then expelled using magnetic forces to create thrust.
ERV	Earth Return Vehicle	Vehicle to carry humans from Mars to Earth.
ESA	European Space Agency	Space agency of Europe, comprised of 20 member nations.
EVA	Extra-Vehicular Activity	When an astronaut goes outside a vehicle (e.g. spacecraft or rover) or habitat wearing a spacesuit.
FMARS	Flashline Mars Arctic Research Station	MARS at Haughton Crater on Devon Island in the Canadian Arctic.
G-FOLD	Guidance for Fuel Optimal Large	Navigation system that calculates optimised trajectory

	Diverts	corrections in real time.
GNC	Guidance, Navigation and Control	Spacecraft subsystem responsible for controlling its position and velocity.
GPS	Global Positioning System	Navigation system that can pinpoint your location anywhere on Earth.
GRS	Gamma Ray Spectrometer	An instrument on Mars Odyssey.
HD	High Definition	Camera, image or video with high resolution.
HEEO	Highly Elliptical Earth Orbit	An equatorial orbit around Earth with a low perigee (~1000 km) and a high apogee (geostationary altitude or higher).
HEMO	Highly Elliptical Mars Orbit	An equatorial orbit around Mars with a low periareon and a high apoareon.
HiRISE	High Resolution Imaging Science Experiment	Camera in Mars orbit sending back beautiful high resolution colour images of Mars.
HI-SEAS	Hawaii Space Exploration Analog and Simulation	A MARS on the slopes of the Mauna Loa volcano in Hawaii.
HLCRS	Heliocentric Laser Communications Relay Satellite	An LCRS in orbit around the Sun.
HLLV	Heavy Lift Launch Vehicle	Rocket capable of lifting between 20 and 50 tonnes to LEO.
HMM	Human Mars Mission	Space mission that sends humans to Mars.
HSF	Human Space Flight	Any activities involving humans in space.
IMRS	International Mars Research Station	Mars base to be developed by the world's leading space agencies and made available to all nations for science and engineering research.
IP	Internet Protocol	Communications protocol developed for the Internet.
ISAP	In Situ Air Production	Producing breathable air from local resources.
ISECG	International Space Exploration Coordinating Group	International consortium of space agencies collaborating on global space exploration policies and strategies.
ISEP	In Situ Electricity Production	Producing electrical energy from local resources.
ISFP	In Situ Food Production	Producing food from local resources.

ISPP	In Situ Propellant Production	Producing rocket propellant from local resources.
ISRO	Indian Space Research Organisation	Space agency of India.
ISRU	In Situ Resource Utilisation	Making use of local resources.
ISS	International Space Station	Space station built by international partners.
ISWP	In Situ Water Production	Producing potable water from local resources.
IT	Information Technology	Software, hardware and networking.
ITN	Interplanetary Transport Network	A collection of gravitationally-determined pathways through the Solar System that require minimum energy for a spacecraft to follow.
IVA	Intra-Vehicular Activity	Activity inside a vehicle (e.g. spacecraft or rover) or habitat.
JAXA	Japan Aerospace Exploration Agency	Space agency of Japan.
JSA	Job Safety Analysis	A process of optimising safety for a work task (such as an EVA) by going through the steps involved, identifying possible hazards, and developing strategies for mitigating these.
KARI	Korea Aerospace Research Institute	Space agency of South Korea.
LCRD	Laser Communications Relay Demonstration	Mission to demonstrate laser communications between satellites, ground stations and spacecraft, throughout Solar System.
LCH4	Liquid Methane	Versatile and inexpensive rocket fuel.
LCRS	Laser Communications Relay Satellite	A CRS utilising laser communications.
LED	Light-Emitting Diode	Diode that produces coloured light when current is flowing through it, commonly used in consumer electronics.
LEO	Low Earth Orbit	Orbit around Earth that is low (e.g. the ISS is in LEO at about 300 km up).
LH2	Liquid Hydrogen	Hydrogen in liquid form, as used in rocket propellant.
LION	Landing with Inertial and Optical Navigation	GNC technology that uses landmark recognition to facilitate pinpoint landings on worlds without positioning satellites.
LLCD	Lunar Laser Communication	Experiment in laser communications between a satellite in lunar orbit (Lunar Atmosphere and Dust Environment

	Demonstration	Explorer) and a ground station on Earth.
LNG	Liquid Natural Gas	Mined gaseous hydrocarbon fossil fuel, comprised mainly of methane.
LMO	Low Mars Orbit	Orbit around Mars that is low (e.g. 200 km).
LOC	Loss Of Crew	Something bad that could happen during a mission.
LOM	Loss Of Mission	Not as bad as LOC, but still fairly undesirable.
LOX	Liquid Oxygen	Oxygen in liquid form, as used in rocket propellant.
LPS	Local Positioning System	A navigation system that covers only a small area of land (c.f. GPS), usually defined by three or more signalling beacons.
LZ	Landing Zone	Patch of open ground where an aerial or space vehicle can land.
MAC	Mars Ascent Capsule	Crew Dragon which forms the topmost section of the MAV. Also known as *Kepler*.
MARS	Mars Analogue Research Station	Base in a Mars analogue environment on Earth where simulated Mars missions are conducted.
MAV	Mars Ascent Vehicle	Vehicle to carry people from Mars surface to Mars orbit.
MCC	Mission Control Centre	Team and facility on Earth that manage a Mars mission and communicate with the crew.
MCOS	Mars Communication and Observation Satellite	Satellite in aerostationary orbit with the same longitude as the IMRS, for maintaining communications link with Earth and watching the base.
MCP	Mechanical Counter-Pressure	Type of spacesuit that uses elasticised material instead of gas to provide pressure to an astronaut's body.
MCT	Mars Colonial Transporter	A new SHLLV being developed by SpaceX, capable of delivering up to 100 tonnes to the surface of Mars.
MDRS	Mars Desert Research Station	MARS near Hanksville in Utah, USA.
MER	Mars Exploration Rover	Twin rovers *Spirit* and *Opportunity* that have spent many years exploring Mars.
MGS	Mars Global Surveyor	NASA spacecraft in Mars orbit.
MIT	Massachusetts Institute of Technology	University in Massachusetts, USA, known for developing breakthrough technologies.
MLI	Multi-Layer Insulation	Multiple layers of thin sheets of material designed to reduce heat loss by thermal radiation.
MLLV	Medium Lift Launch Vehicle	Rocket capable of lifting between 2 and 20 tonnes to LEO.
	Mars Orbit	Orbital manoeuvre that moves a spacecraft into Mars

MOI	Insertion	orbit.
MOLA	Mars Orbiting Laser Altimeter	Instrument onboard MGS that measures the topography of Mars.
MOR	Mars Orbit Rendezvous	When two spacecraft meet and dock in Mars orbit.
MSA	Mars Society Australia	Australian branch of the Mars Society.
MSC	Mars Supply Capsule	Cargo Dragon used to deliver supplies to the Martian surface.
MTO	Mars Transfer Orbit	A Hohmann transfer orbit that connects Earth's orbit with Mars's.
MTV	Mars Transfer Vehicle	Vehicle to carry humans from Earth to Mars.
NASA	National Aeronautics and Space Administration	Space agency of the USA.
NDS	NASA Docking System	Universal docking system used on the ISS, Soyuz, Dragon, B330 and other spacecraft.
NEP	Nuclear Electric Propulsion	Form of EP where the electricity is generated from nuclear energy.
NERVA	Nuclear Engine for Rocket Vehicle Application	An American program to develop a thermal nuclear propulsion system for interplanetary crewed missions.
NTR	Nuclear Thermal Rocket	Rocket with a propulsion system that uses nuclear fission to superheat reaction mass.
OMS	Orbital Manoeuvring System	System of thrusters for manoeuvring on orbit.
OPALS	Optical PAyload for Lasercomm Science	Experiment in laser communications between ISS and ground stations.
PHP-M	Permanent Human Presence on Mars	When there's always one or more people on Mars.
PLSS	Personal Life Support System	Part of a spacesuit that maintains a healthy environment for the occupant.
PT	Physical Training	Exercise.
PV	Photovoltaic	Conversion of light into electricity.
RCS	Reaction Control System	System of thrusters that control a spacecraft's orientation in space.
RH	Relative Humidity	Amount of water vapour in the atmosphere expressed as a degree of saturation.

RLS	Reusable Launch System	Rocket that can be reused multiple times.
ROI	Return On Investment	What you get back from an investment in a project or business.
RP-1	Rocket Propellant-1 *or* Refined Petroleum-1	Highly refined form of kerosene used as rocket fuel.
RTG	Radioisotope Thermoelectric Generator	Generator that converts heat produced by radioactive decay into electricity.
RWGS	Reverse Water Gas Shift	Chemical reaction that reacts CO_2 with H_2 to produce CO and H_2O.
SEP	Solar Electric Propulsion	Form of EP where the electricity is generated from solar energy.
SCUBA	Self-Contained Underwater Breathing Apparatus	Equipment to enable breathing when diving underwater.
SHAB	Mars Surface Habitat	Habitat for the surface of Mars.
SHLLV	Super Heavy Lift Launch Vehicle	Rocket capable of lifting more than 50 tonnes to LEO.
SLS	Space Launch System	Space Shuttle-derived SHLLV currently in development by NASA.
SN	Space Network	Network of satellites and ground stations that support communications with spacecraft close to Earth.
SOEC	Solid Oxide Electrolysis Cell	A solid oxide fuel cell operated in regenerative mode to separate CO_2 into CO and O_2.
SP	Special Publication	Publication that is special (esp. NASA).
SSAU	State Space Agency of Ukraine	Space agency of Ukraine.
SSTO	Single Stage To Orbit	Rocket that can reach orbit with only one stage.
STEM	Science, Technology, Engineering and Mathematics	Major educational topics related to space exploration and settlement.
TCP	Transmission Control Protocol	Internet protocol that ensures reliable data communication.
TDRSS	Tracking and Data Relay Satellite System	Network of communications satellites and ground stations used by NASA for space communications.
TEI	Trans-Earth Injection	Orbital manoeuvre that places a spacecraft on a trajectory towards Earth.

THAB	Mars Transit Habitat	Habitat for the flight from Earth to Mars and back.
TMI	Trans-Mars Injection	Orbital manoeuvre that places a spacecraft on a trajectory towards Mars (see MTO).
TRL	Technology Readiness Level	Level of maturity of a technology, and an assessment of its suitability for space applications.
TWR	Thrust-to-Weight Ratio	The ratio between an engine's thrust and its weight in Earth gravity.
VASIMR	Variable Specific Impulse Magnetoplasma Rocket	Spacecraft propulsion technology that uses radio waves to to produce a plasma, which is then accelerate using magnetic fields.
VTOL	Vertical Take-Off and Landing	Refers to aircraft or spacecraft that can take-off, hover and land vertically.
WAVAR	WAter Vapour Adsorption Reactor	Device for extracting water from the Martian atmosphere.
WRS	Water Recovery System	A high-efficiency water recycling unit for a space habitat.
ZPB	Zero Prebreathe	When no time is required for prebreathing before EVA, usually because of the spacesuit technology.
ZPS	Zero Prebreathe Spacesuit	Spacesuit providing a high enough pressure that prebreathing is not required.

References

1. B. Acikmese, J. Casoliva and J. M. Carson III, "G-FOLD: A Real-Time Implementable Fuel Optimal Large Divert Guidance Algorithm for Planetary Pinpoint Landing," Concepts and Approaches for Mars Exploration, 2012.

2. S. Adan-Plaza, K. Carpenter, L. Elias, R. Grover, M. Hilstad, C. Hoffman, M. Schneider and A. Bruckner, "Extraction of Atmospheric Water on Mars for the Mars Reference Mission," HEDS-UP Mars Exploration Forum, LPI Contribution No. 955, May 1998.

3. J. Bauer, S. Hengsbach, I. Tesari, R. Schwaiger and O. Kraft, "High-strength cellular ceramic composites with 3D microarchitecture", Proc. Natl. Acad. Sci. USA, vol. 111 no. 7, 2453-2458, Jan 2014.

4. G. Bonin, "Reaching Mars for Less: The Reference Mission Design of the MarsDrive Consortium," 25th International Space Development Conference, Los Angeles, CA, May 2006.

5. D. Byford, J. Goppert and T. Gangale, "Optimal Location of Relay Satellites for Continuous Communication With Mars," AIAA 2008-7919. Presented at Space 2008, 9 Sep 2008, San Diego, CA.

6. P. D. Campbell, "Internal Atmospheric Pressure and Composition for Planet Surface Habitats and Extravehicular Mobility Units," Lockheed Engineering and Sciences Company, Contract NAS9-17900, Job Order K1-ETB, Report No. JSC-25003, for NASA Man-Systems Division, 1991.

7. P. R. Christensen, J. L. Bandfield, V. E. Hamilton, S. W. Ruff, H. H. Kieffer, T. N. Titus, M. C. Malin, R. V. Morris, M. D. Lane, R. L. Clark, B. M. Jakosky, M. T. Mellon, J. C. Pearl, B. J. Conrath, M. D. Smith, R. T. Clancy, R. O. Kuzmin, T. Roush, G. L. Mehall, N. Gorelick, K. Bender, K. Murray, S. Dason, E. Greene, S. Silverman and M. Greenfield, "Mars Global Surveyor Thermal Emission Spectrometer experiment: Investigation description and surface science results", J. Geophys. Res.,106, 23,823-23,871, 2001.

8. M. Cohen, M. T. Flynn, R. L. Matossian, "Water Walls Architecture: Massively Redundant and Highly Reliable Life Support for Long Duration Exploration Missions," Global Space Expl. Conf., Washington D.C., GLEX-2012.10.1.9x12503, 2012.

9. C. Cooper, W. Hofstetter, J. A. Hoffman and E. F. Crawley, "Assessment of architectural options for surface power generation and energy storage on human Mars missions," Acta Astronaut., vol. 66, no. 7–8, pp. 1106–1112, Apr 2010.

10. J. Delaune, G. Le Besnerais, M. Sanfourche, T. Voirin, C. Bourdarias and J. Farges, "Optical Terrain Navigation for Pinpoint Landing: Image Scale and Position-Guided Landmark Matching," Proceedings of the 35th Annual Guidance and Control Conference, 2012.

11. J. L. Dickson, C. I. Fassett and J. W. Head, "Amazonian-aged fluvial valley systems in a climatic microenvironment on Mars: Melting of ice deposits on the interior of Lyot Crater", GRL, vol. 36, 2009.

12. B. G. Drake, "Human exploration of Mars, Design Reference Architecture 5.0", July 2009, p. 100., Mars Architecture Steering Group, NASA/SP–2009–566.

13. E. C. Ethridge and W. F. Kaukler, "Microwave Extraction of Volatiles for Mars Science and ISRU," AIAA Aerospace Sciences Meeting, 2012.

14. M. J. Fogg, "The Utility of Geothermal Energy on Mars," J. Br. Interplanet. Soc., vol. 49,

pp. 403–422, 1996.

15. D. Gage, "Mars Base First: A Program-level Optimization for Human Mars Exploration", J. Cosmol., vol 12, pp. 3904-3911, 2010.

16. T. Gangale, "MarsSat: Assured Communication With Mars," Ann. NY Acad. Sci., vol. 1065, 1 Jan 2006. Presented at New Trends in Astrodynamics and Applications II, 3 June 2005, Princeton U., Princeton, NJ.

17. G. D. Grayson, M. L. Hand, E. C. Cady, 2009, "Thermally coupled liquid oxygen and liquid methane storage vessel", US Patent 7568352 B2.

18. M. R. Grover, M. O. Hilstad, L. M. Elias, K. G. C. M. A. Schneider, C. S. Hoffman, S. Adan-Plaza and A. P. Bruckner, "Extraction of Atmospheric Water on Mars in Support of the Mars Reference Mission," MAR 98-062, Proceedings of the Founding Convention of the Mars Society: Part II, ed. R. M. Zubrin and M. Zubrin, Boulder, CO, pp. 659-679, August 13-16, 1998.

19. M. R. Grover, E. Sklyanskiy, A. D. Steltzner and B. Sherwood, "Red Dragon-MSL Hybrid Landing Architecture for 2018," JPL, Caltech, Concepts and Approaches for Mars Exploration, 2012.

20. A. J. Hanford (ed.), "Advanced Life Support Baseline Values and Assumptions Document," NASA JSC, 2004.

21. M. A. Interbartolo III, G. B. Sanders, L. Oryshchyn, K. Lee, H. Vaccaro, E. Santiago-Maldonado and Anthony C. Muscatello, "Prototype Development of an Integrated Mars Atmosphere and Soil-Processing System," J. Aerosp. Eng., vol 26, SPECIAL ISSUE: In Situ Resource Utilization, pp. 57–66, 2013.

22. JSC 20584, "Spacecraft Maximum Allowable Concentrations For Airborne Contaminants," Toxicology Group, Medical Operations Branch, Medical Sciences Division, Space and Life Sciences Directorate, NASA JSC, June 1999.

23. J. S. Karcz, S. M. Davis, M. J. Aftosmis, G. A. Allen, N. M. Bakhtian, A. A. Dyakonov, K. T. Edquist, B. J. Glass, A. A. Gonzales, J. L. Heldmann, L. G. Lemke, M. M. Marinova, C. P. Mckay, C. R. Stoker, P. D. Wooster and K. A. Zarchi, "Red Dragon: Low-Cost Access to the Surface of Mars Using Commercial Capabilities," Concepts and Approaches for Mars Exploration, 2012.

24. J. Kozicki and J. Kozicka, "Human friendly architectural design for a small Martian base," Adv. Sp. Res., vol. 48, no. 12, Dec 2011.

25. M. A. Kreslavsky and J. W. Head III, Mars: Nature and evolution of young latitude-dependent water-ice-rich mantle, Geophys. Res. Letters, 29, 2002.

26. K. E. Lange, C. H. Lin, B. E. Duffield and A. J. Hanford, "Advanced Life Support Requirements Document," JSC-38571, Revision C, NASA JSC, Houston, Texas, 2003.

27. B. Lansdorp and A. A. Wielders, "Mars One Communications System," http://www.mars-one.com/technology/communications-system (Retrieved 2014-05-26).

28. W. J. Larson and L. K. Pranke, "Human Spaceflight: Mission Analysis and Design (Space Technology Series)," McGraw-Hill, 1999.

29. A. LeBlanc, T. Matsumoto, J. Jones, J. Shapiro, T. Lang, S. M. Smith, L. Shackelford, J. Sibonga, H. Evans, E. Spector, T. Nakamura, K. Kohri and H. Ohshima, "Bisphosphonate as a Countermeasure to Space Flight-Induced Bone Loss," American Society for Bone and Mineral Research, Houston, TX, JSC-CN-30935, 2014.

30. M. Raftery, D. Cooke, J. Hopkins and B. Hufenbach, "An Affordable Mission to Mars," 64th International Astronautical Congress, Beijing, China, IAC-13, A5, 4-D2.8.4, 2013.

31. R. Rhinehart, "Soylent - Free Your Body," https://campaign.soylent.me/soylent-free-your-body (Retrieved 2013-10-15).

32. Peter B. de Selding (2010-02-03), "ESA Chief Lauds Renewed U.S. Commitment to Space Station, Earth Science", http://spacenews.com/esa-chief-lauds-renewed-us-commitment-space-station-earth-science (Retrieved 2015-01-26).

33. C. R. Stoker, A. Davila, S. Davis, B. Glass, A. Gonzales, J. Heldmann, J. Karcz, L. Lemke and G. Sanders, "Ice Dragon: A Mission to Address Science and Human Exploration Objectives on Mars," Concepts and Approaches for Mars Exploration, 2012.

34. G. W. W. Wamelink, "Growing plants on Mars: Wageningen UR goes extraterrestrial", Wageningen UR, https://www.wageningenur.nl/en/show/Growing-plants-on-Mars-Wageningen-UR-goes-extraterrestrial.htm, 2013 (Retrieved 2014-06-09).

35. A. Wielders, B. Lansdorp, S. Flinkenflögel, B. Versteeg, N. Kraft, E. Vaandrager, M. Wagensveld, A. Dogra, B. Casagrande and N. Aziz, "Mars One: Creating a human settlement on Mars," European Planetary Science Congress 2013, vol. 8, 2013.

36. D. Willson and J. D. A. Clarke, "A Practical Architecture for Exploration-Focused Manned Mars Missions Using Chemical Propulsion, Solar Power Generation and In-Situ Resource Utilisation," J. Br. Interplanet. Soc., vol. 58, pp. 181–196, 2005.

37. R. M. Zubrin, "The Case for Colonizing Mars", Ad Astra, July/August 1996, http://www.nss.org/settlement/mars/zubrin-colonize.html (Retrieved 2015-01-29).

38. R. M. Zubrin, D. A. Baker and O. Gwynne, "Mars direct - A simple, robust, and cost effective architecture for the Space Exploration Initiative," 29th Aerosp. Sci. Meet. AIAA, 1991.

39. R. M. Zubrin w/ R. C. Wagner, 1996, "The Case for Mars: The Plan to Settle the Red Planet and Why We Must," ISBN 0-684-83550-9 978-0684835501, Simon & Schuster, NY.

Thanks!

Thanks for reading my humble book :) I hope you enjoyed it.

Please send comments, feedback, suggestions and corrections to shaun@astromultimedia.com.

I also hope you will feel inspired to help with ongoing development of this dream of colonising another world. You may like to join the Mars Society, Mars Settlement Research Organisation, Explore Mars, or any of the other great space, Moon and Mars communities, where you'll have the opportunity to connect with like-minded visionaries of all ages.

Ultimately we are all brothers and sisters on this truly noble, spiritual, evolutionary and revolutionary quest to expand into space and develop this exciting new chapter of human history. There are few things we can do that will produce a greater benefit for our species and Earth. And we will do it.

Together, we can make humanity multiplanetary.

The author with Athena concept sketch (Credit: Rob Wibaux)

Have a great day! Or, if you're reading this on Mars - have a great sol!

Shaun Moss

March 2015

= From Mars to the Stars =